D1179441

The Design of Forest Landscapes

HOUGHALL
COLLEGE
LIBRARY

HOUGHALL COLLEGE LIBRARY	
CLASS No.	715 · 2
ACC No.	95334
ORDER No.	C147

The Design of Forest Landscapes

Oliver W. R. Lucas
Forestry Commission

OXFORD UNIVERSITY PRESS
1991

Oxford University Press, Walton Street, Oxford OX2 6DP
Oxford New York Toronto
Delhi Bombay Calcutta Madras Karachi
Petaling Jaya Singapore Hong Kong Tokyo
Nairobi Dar es Salaam Cape Town
Melbourne Auckland
and associated companies in
Berlin Ibadan

Oxford is a trade mark of Oxford University Press

Published in the United States
by Oxford University Press, New York

© *Oliver W. R. Lucas 1991*

All rights reserved. No part of this publication may be reproduced,
stored in a retrieval system, or transmitted, in any form or by any means,
electronic, mechanical, photocopying, recording, or otherwise, without
the prior permission of Oxford University Press

This book is sold subject to the condition that it shall not, by way
of trade or otherwise, be lent, re-sold, hired out, or otherwise circulated
without the publisher's prior consent in any form of binding or cover
other than that in which it is published and without a similar condition
including this condition being imposed on the subsequent purchaser

British Library Cataloguing in Publication Data
Lucas, Oliver W. R.
  The design of forest landscapes.
  1. Great Britain. Forests. Landscape design
  I. Title
  715.2
  ISBN 0−19−854280−1

Library of Congress Cataloging-in-Publication Data
Lucas, Oliver W. R.
  The design of forest landscapes / Oliver W. R. Lucas.
    p. cm.
Includes bibliographical references and index.
  1. Forest landscape design. I. Title.
  SB475.9.F67L83 1990
  634.9′5—dc20 90-45536
                    CIP
ISBN 0−19−854280−1

Printed and bound in Hong Kong

ERRATUM

On p. iv, the copyright line should read
© Crown copyright 1991. Published by per-
mission of the Controller of Her Majesty's
Stationery Office

Foreword by
Dame Sylvia Crowe DBE PPILA

Landscape Consultant to the Forestry Commission 1963–75

Since the author of this book is working within the Forestry Commission, he is able to speak with authority on both the possibilities and the problems of reconciling commercial forestry with conservation, good landscape, and public enjoyment.

His technical knowledge of landscape problems should give valuable guidance to foresters, and encourage a better understanding between them and the general public, to whom the forests now represent a major recreational asset.

The landscape of Britain has suffered deforestation for centuries, and now that some increase of woodland is under way it is vital that it should be carried out with sympathy for both the visual landscape and the native woodland community of plants, birds, and animals.

This has been a major objective in the Forestry Commission's work for many years past and the publication of this book is a further step in this direction. It should be of value both to Commission staff and to those working in the field of private forestry.

Acknowledgements

The principles and practice described here were developed over the years by a number of people. The initial advocacy and principles of Dame Sylvia Crowe were extensively developed by Duncan Campbell into practice which made forest design substantially more accessible to foresters. He also initiated and advised on the content of this book. Much of the practice described has been developed and tested with great enthusiasm and tolerance by Forestry Commission foresters too numerous to mention.

The illustrations have been prepared by Anne Baxter and Elaine Dick from the roughest sketches. Bringing the manuscript to publication has been a major effort, achieved as a result of Alastair Rowan's editing and Simon Bell's persistence.

Thank you all.

O. W. R. L.

Ordnance Survey Acknowledgements

Fig. 2.17 Reproduced from the Ordnance Survey 1:10000 maps NN51NE (1978) and NN51NW (1978), with the permission of the Controller of Her Majesty's Stationery Office, ©, Crown Copyright.

Fig. 4.10 Reproduced from the Ordnance Survey map Sheet 32, 1:50000 *South Skye* (1976), with the permission of the Controller of Her Majesty's Stationery Office, ©, Crown Copyright.

Fig. 9.7(g) Reproduced from the Ordnance Survey Sheet 78, 1:50000 *Moffat* (1976), with the permission of the Controller of Her Majesty's Stationery Office, ©, Crown Copyright.

Fig. 11.8 Reproduced from the Ordnance Survey 1:10000 map NS63NE (1980), with the permission of the Controller of Her Majesty's Stationery Office, ©, Crown Copyright.

Fig. 12.17 Reproduced from the Ordnance Survey 1:10000 maps SH54NE (1977), SH54NW (1978), SH55SE (1978), and SH55SW (1978) with the permission of the Controller of Her Majesty's Stationery Office, ©, Crown Copyright.

Fig. 16.5 Reproduced from the Ordnance Survey 1:10000 map SH54NW (1978), with the permission of the Controller of Her Majesty's Stationery Office, ©, Crown Copyright.

Contents

10 Change of species   **189**

11 Visual impact of forest operations   **204**

12 Felling and restocking   **225**

Introduction

The nature of the landscape is determined by the natural qualities of geology and climate, by the changes made by man's use of the land and by our responses to what we see. Man's widespread influence over the contrasting patterns of woodland and open ground has increased through a long history of agricultural development, forest clearance, and, more recently, afforestation. As the landscape has gradually become more artificial in its appearance so has the demand increased for more natural surroundings as a contrast to an increasingly urban, crowded, and stressful existence. Woodlands and forests must, therefore, be an attractive and harmonious feature of our countryside while meeting the needs of society by producing timber, creating varied and extensive wildlife habitats, and providing a place for recreation. Landscape design aims to achieve these material needs and maintain our pleasure in the landscape.

The natural forests which colonized most of Britain after the last ice age were cleared or drastically altered by man with the development of agriculture. Grazing animals prevented the regeneration of the forests and increasing requirements of timber for industrial purposes caused the forest area to diminish to a scant 83 000 hectares (less than 6 per cent of the lands surface of Britain) by the end of the nineteenth century. Critical shortages of timber in the 1914–18 war led to a policy of woodland expansion and the Forestry Commission was set up as the government department with responsibility for carrying out this policy. The new forests established by the Forestry Commission were usually in the uplands where land was more readily available. Upland afforestation presented severe technical problems up to the early years of this century, but these were overcome by successful methods of draining and cultivation, and the introduction of conifers hardy enough to flourish on the impoverished soils and inhospitable conditions of upland Britain.

The area of privately owned woodlands in the uplands has also greatly increased, notably over the last 25 years.

The rapid expansion of forests and the need to create a strategic reserve of timber led to many areas being planted as quickly as possible. Boundaries often followed the geometric enclosures of the eighteenth or nineteenth centuries, with straight fence-lines. The forest was often on a much larger scale than the existing enclosure patterns and the contrast with open hill often created an intrusive appearance, with artificial shapes appearing to conflict with rounded landforms and the natural irregularity of vegetation, rocks, and watercourses. Where planting had been done with a range of species following patterns of ground vegetation, some attractive forest landscapes were created, but it is now accepted that errors of landscape design were often made in the layout of the early twentieth century British forests.

Criticism of the appearance of these forests was inevitable, particularly by those who used the uplands for recreation. For many years the arguments for and against the expansion of forestry were confounded by strongly polarized views and confusion of ecological issues with perception of the landscape. Those concerned about the reduction of open space in the countryside have to recognize that the re-establishment of Britain's forests is a legitimate response to the needs of society; while foresters must recognize proper and reasonable desires to conserve open lands for aesthetic and wildlife conservation reasons.

The Forestry Commission was conscious of this issue, and of the necessity for a good standard of forest design. The Commission appointed Dame Sylvia Crowe as its landscape consultant in 1963; she showed how timber-producing forests could be reconciled with the landscape, and established sound principles of forest design which have been applied and developed by her successors, and by Forestry Commission landscape architects. In 1980 the Commission included protection and enhancement of the environment as one of its objectives, and in 1985 the Forestry Act 1967 was amended to give statutory force to this aim.

Private foresters have also become conscious of the need to achieve a good standard of landscape design, and this is reflected in the *Forestry and Woodland Code* published by Timber Growers United Kingdom in 1986. The role of the Forestry Commission in ensuring that private forestry achieves satisfactory standards of landscape design was given emphasis in the Woodland Grant Scheme introduced in 1988. One of its objectives is 'to enhance the landscape, to create new wildlife habitats and to provide for recreation and sporting uses in the longer term'.

Definitions

The *Oxford English Dictionary* defines **landscape** as 'a prospect of inland scenery such as can be taken in at a glance from one point of view'. Scenery is defined as 'the general appearance of a place and its natural features from a picturesque point of view'. Man's observation is a vital aspect of this definition, and the landscape is more than an area of land with its individual arrangement of features. It is also our vision of that area, which, in turn, is influenced by our natural instincts for survival, our emotions, education, culture, and experience.

The landscape can be defined in terms of natural components, human attributes, and aesthetic qualities. Of the natural components, landform and vegetation usually have the greatest influence, with rocks and water in a subsidiary role. The management of vegetation, e.g. hedgerows and woodland, is the most widespread human attribute, with buildings and structures on a more local scale. Aesthetic factors are concerned with the reaction of the mind to what the eye sees. For example, shape and scale have most influence on our personal reactions, and on whether woodland seems in harmony which its surroundings. These factors can be used both to analyse the landscape and to guide the designer.

While it might be argued that the aesthetic factors affecting a landscape are purely a matter of personal taste, the weight of evidence suggests that a consensus exists. Were this not so, urban industrial landscapes would be as highly valued as the English Lake District or the Yorkshire Dales. There is a base level of landscape quality which is reached before personal preference comes into play, for example, in the choice between the rugged grandeur of Glencoe or the gentler landscape of the Scottish Borders.

Objective terms are needed in which to discuss aesthetic matters. Assertions on the form of 'I like' or 'I do not like' neither identify problems nor suggest solutions and bring an end to constructive discussion. Scientifically validated studies into perception, landscape evaluation, and preference are in their infancy and do not always assist design. Previous experience in designed landscapes and other creative activities, such as painting, sculpture, and architecture can, however, provide valid insights and objectives for designs.

The aesthetic terms used in this book are described in the glossary and have been derived from *Visual design* by Lillian Garrett (1967). These terms are useful because they describe both how we perceive things and how we should design for good results.

By definition, the landscape is seen in perspective and this is the only way to assess the visual qualities of a design. A **landscape plan** should, therefore, include accurately co-ordinated perspectives and maps. The design of landscape is complex, and to try to express design principles directly on a map will not work. The view must be seen in perspective illustration to ensure that all the aesthetic factors are properly balanced, that proposed changes are an improvement and that their cost is justified.

The designer soon realizes that the single viewpoint does not take account of people moving around the landscape and that it is more three-dimensional than the above definitions suggest. Memory affects our view of the landscape; as we move about, individual places are seen as continuations or contrasts with those seen before. Our knowledge of the area immediately around the view, but outside the limits of our vision also affects strongly our perception of the landscape.

In its broadest sense **landscape design** is the organization of a place in a way which reconciles the conflicting requirements of use, e.g. forestry, agriculture, wildlife, and recreation, while ensuring an attractive appearance. Since effective design requires both the suppression of ugliness and the enhancement of beauty, both a basic level of design to avoid ugliness everywhere and the inspiration to achieve outstanding appearance in key areas are needed. Such a landscape must not be simply derived from an arbitrary set of aesthetic rules, but from the natural and unique attributes of each site. This means that the design of the forest is influenced by factors beyond its boundaries both in the immediately surrounding landscape and in the wider region.

The merit of a landscape design is less in the nature of the components than in the way they are put together. The components of the forest landscape are mainly the working parts, such as stands of trees for timber and wildlife, roads, felled areas, open space for protection of water quality and for deer control, car parks for visitors, and fences and gates. The main objective of forest landscape design is the shaping of the working parts to achieve a satisfactory appearance that reflects the essence of the site.

The infinite possibilities for variation in design and the unique qualities of individual places mean that comparisons of value can only be made in a qualitative way. The judgement of the value of individual areas is termed **landscape sensitivity** and is based on a careful assessment of aesthetic quality. Advice on this may be obtained from professionally qualified members of the Landscape Institute. Many landscapes of the highest quality have been recognized by statutory designation, e.g. National Parks or National Scenic Areas, and must be protected by applying high standards of forest design. The numbers of local residents, visitors, and volume of traffic all give an indication of the exposure of an area to the public. The greater the number of people, the more sensitive is the landscape.

Economic aspects of landscape

It is difficult to attribute a cash value to landscape. It is enjoyed or disliked without being bought or sold, yet it may well have an effect on the attractiveness of an area from the point of view of tourism. Landscape should, therefore, be seen as a resource to be carefully husbanded. It is easy to establish agreement that certain landscapes are valued more highly than others, and that some landscapes are so unacceptable that they are clearly detrimental to society; it is much more difficult to determine the point where a landscape design becomes marginally acceptable, with no marked positive or negative value. This is the judgement which the Forestry Commission is required to make in carrying out its duty to endeavour to achieve a reasonable balance between the needs of forestry and those of the environment, whether on the forests under its own management or in the administration of private forestry. The basic level of acceptable landscape practice as described in this book is a key element in achieving this balance.

The loss of potential timber production or additional costs due to landscape design can be calculated, but it is difficult for the forest manager to assess the value of individual landscapes and to judge what cost of landscape design is justified. Landscapes of greater sensitivity may require much higher levels of design input if forest operations are to take place without intrusion. The level of resources to be allotted should be assessed for the complete treatment for specific forest areas, reflecting the level of sensitivity, and not by setting arbitrary maximum limits for the site. It is important to look at the opportunity costs of environmental work over the whole of a particular forest enterprise. This helps to indicate how the larger resources allocated to the few outstanding landscapes can be counterbalanced by the areas where the basic level of design is adequate and expenditure is lower. The economic targets set should be reviewed regularly in terms of both visual achievement and cost control.

Attitudes

The principles and practical advice given in this book can be put into practice successfully only if the the right attitudes are adopted by the forest designer. There is no single 'correct' design, but a range of good and bad ones. The designer must stimulate the managers' enthusiasm and support, as well as carrying out the design. To achieve even modest results the designer must pursue aesthetic excellence persistently, because forest design is vulnerable to uncertainties such as windthrow and the varying demands of timber markets. It is always easier to achieve a modest standard by reduction from the excellent than to try to improve on the inadequate. The designer must be ruthlessly critical of his own work, and seek and accept constructive criticism in an objective spirit. He should keep an open mind to new aspects or alternatives to a design until a solution is agreed by the client or his authorized manager. In many cases, it will be necessary for the designer to provide advice over a period of time, during which the original design may have to be modified to take account of changed circumstances, such as windthrow or markets.

This book is intended to provide managers of woods and forests with an outline of the principles of forest landscape design and advice on how they may be put into practice. It should also be a useful reference for landscape architects and planners involved in forestry.

Landscape design is a complex subject which requires an understaning of basic principles. These are set out in the early chapters and practitioners should be familiar with them before referring to the practical applications given later. There are no quick and easy recipes for successful design.

Design principles

Landscape encompasses the features and materials present on an area of land, and our reaction to them. Our natural behaviour and previous experience affect how we react to what we see; the mind is involved as well as the eye. This combination of mind and eye is known as perception, and this chapter explains how an understanding of perception is basic to the design of good forest and woodland landscapes.

There is no formula for easy success, but by using the six most important principles outlined in this chapter the worst mistakes can be avoided and ways found to make improvements. If these principles are well understood, applying the advice in the rest of the book will be easier and more effective.

The range of design terms is such that a degree of organization will help to understand them. Design terms can be classified into basic elements, variables, and organizational factors.

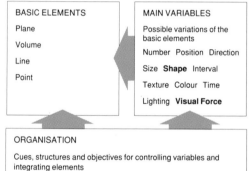

BASIC ELEMENTS

Plane

Volume

Line

Point

MAIN VARIABLES

Possible variations of the basic elements

Number Position Direction

Size **Shape** Interval

Texture Colour Time

Lighting **Visual Force**

ORGANISATION

Cues, structures and objectives for controlling variables and integrating elements

Grouping: Nearness Closure Similarity

Structure: Rhythm Tension Balance **Scale**

Objectives: **Unity Diversity Genius Loci**

Fig. 2.1. The relationship between aesthetic factors. Individual terms are defined in the glossary and those highlighted are explained in detail in this chapter.

What we see around us can be described in terms of four types of basic elements or 'building blocks': volume (three-dimensional), plane (two-dimensional), line (one-dimensional), and point (which has significant position, but almost no dimensions). These elements can vary in a number of ways (variables), often more than one at a time.

A collection of planes, for example, may vary in size, shape, number, position, direction, interval, texture, colour, visual force, and with time and light. The variety of possible combinations is huge. If an overall effect is to be produced, ways of organizing elements and variables are needed; these are known as organizational factors.

Organizational factors can be used to group elements, e.g. nearness, while others such as rhythm, balance, scale, and tension are used to give overall structure to a design.

Shape, visual force, scale, diversity, unity, and the spirit of the place or *genius loci* are the most important principles in forest landscape design. They have the greatest impact on perception and have been found to be closely related to good or bad results. In the examination of these principles which follows some abstract and architectural examples are given because they often make the ideas clearer.

All principles interact in complex and varied ways, and overall objectives are required to achieve satisfactory results. Combining diversity with unity, and the conservation and enhancement of *genius loci* are important objectives.

1. Shape

- Shape has the most powerful and evocative effect on how we see our surroundings.

- Shape is concerned with the edges of planes and volumes (form) and with lines.

- A wide range of shapes can be identified, from the most geometric (artificial) to the most irregular (natural).

- Compatible shapes are most important for overall unity of the landscape.

- Overall shape tends to dominate detail of line and the mind tends to pick out any geometric quality.

- Shape consistently dominates other design factors; incompatible shapes appear intrusive even when scale, diversity, and other factors are well resolved.

Shape is concerned with lines and boundaries; between solid and space, forest and open land, contrasting species, and so on. It can be two- or three-dimensional, and it may describe inner or outer surfaces. Shape describes how different parts of an edge or line relate to one another.

Human perception of shape is so strong that it tends to dominate other visual factors, to the extent that when a shape is taken out of context in terms of scale and position, and expressed in outline only, we identify it readily.

In forest design the most important distinction is between geometric and irregular shapes. Geometric shapes in the landscape are generally an expression of man's impact.

Although some natural forms are geometric, those in natural landscape are usually irregular, diagonal, and asymmetric. The basic geometric shapes are the triangle, square, and circle, and we tend to perceive shapes as distortions, combinations, or variations of them.

Fig. 2.2. Topiary bird at Levens Hall, Cumbria.

Fig. 2.4. Geometric shapes stand out starkly against the natural form of the land in all but the flattest landscapes.

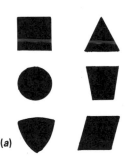

(a)

(b)

Fig. 2.3. (a) Basic geometric shapes. (b) Natural shape.

Compatibility of shape goes further than the difference between geometric and natural. Different geometric shapes may be incompatible, or there may be conflict between one natural shape and another. The closer shapes are to one another, the more compatible they should be.

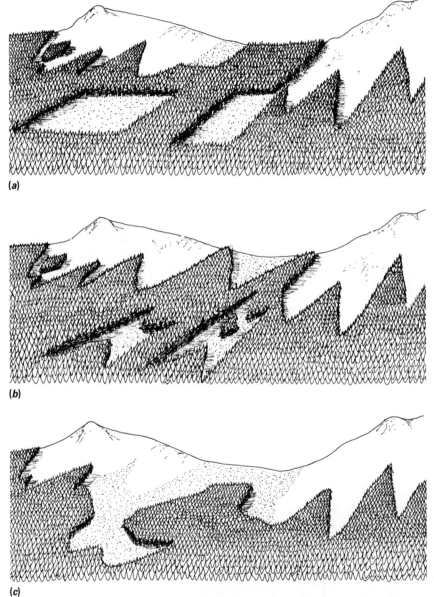

(a)

(b)

(c)

Fig. 2.5. (a) Square felling shapes are incompatible with the spiky upper margin. (b) Angular coupe shapes are more compatible with the upper margin, but scale and symmetry are inappropriate. (c) Increasing the scale, reducing symmetry, and a gentler line in the saddle produces a more unified design, though further improve- ments to the upper margin are still needed.

Fig. 2.6. The geometric lines of the horizontal upper forest edge and the roads are in conflict with the diagonal curving plane of the hillside.

Proportion, direction, and line

Perception of shape is affected by overall proportion and direction, and by the nature of the line that forms the edge. The same basic shape may be expressed by a variety of lines and geometric proportions are readily apparent even when connected by very irregular lines.

A shape derived from asymmetrically positioned points lends itself more readily to a variety of lines which (except for the straight and the zig-zag) give a more or less natural effect.

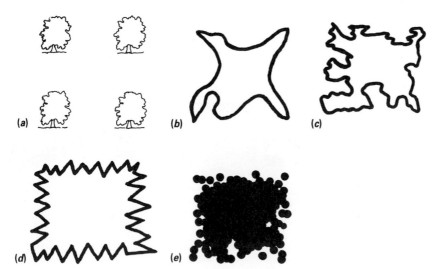

Fig. 2.7. (a) A basic square can be implied by four points: (b) even if the connecting lines are very sinuous or (c) ragged, (d) zig-zag, or (e) diffuse.

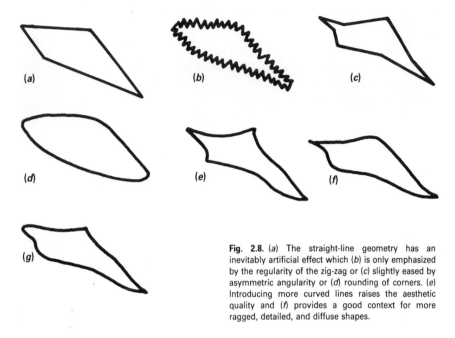

Fig. 2.8. (*a*) The straight-line geometry has an inevitably artificial effect which (*b*) is only emphasized by the regularity of the zig-zag or (*c*) slightly eased by asymmetric angularity or (*d*) rounding of corners. (*e*) Introducing more curved lines raises the aesthetic quality and (*f*) provides a good context for more ragged, detailed, and diffuse shapes.

Position of shapes

Shapes can be positioned so that they are horizontal, diagonal, or vertical. Diagonal shapes give a more dynamic effect than horizontal or vertical shapes, the latter having strong associations with the human figure. As a result diagonal orientation is preferable in landscape, although asymmetric shapes using no parallel edges or right angles can be positioned horizontally. Vertical orientation of

Fig. 2.9. Gable ends of the building reflect rock formations in the slope behind.

shapes, at right angles to the contour, rarely looks satisfactory because the proportions of the landscape are broadly horizontal.

The position and directional qualities of shape can be used to relate geometric to natural forms. Shape can also draw the eye in different directions, a phenomenon which is discussed under the next design principle, visual force.

2. Visual force

- Visual force occurs when a static image gives an illusion of energy or movement.

- The eye and mind respond to visual force in a predictable and dynamic way.

- Visual forces occur in graphic designs, in architecture, and in landscape.

- Visual forces in landform draw the eye down convex slopes and up concave slopes.

- The strength of visual force depends on the scale and irregularity of the landform.

- Forest shapes should be designed to follow visual forces in landform by rising in hollows and falling on spurs in order to create a direct and well unified relationship between the two.

- Natural vegetation shapes can produce similar effects to visual forces in landform in many landscapes.

Visual force is an illusion of movement created by a static image to which the eye and mind respond in a predictable and dynamic way. The more one element responds to the visual forces of another, the more they are perceived as parts of an overall composition, the stronger the visual unity, and the higher the aesthetic quality.

Visual forces in shapes

Visual forces can be created by shapes and these are most apparent in their directional qualities. The arrow is the simplest and most familiar. It is so powerful that a chevron associated with any

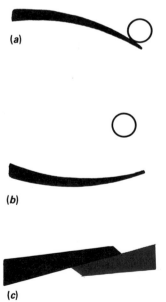

(a)

(b)

(c)

Fig. 2.10. The illustrations suggest (a) a bending downwards and rolling to the right, (b) an upward bounce, and (c) a shearing movement.

Fig. 2.11. The eye is drawn to the end of the corridor by the sinuous lines of the wall, the floor, and the sill.

Fig. 2.12. There is an illusion of movement in both (*a*) and (*b*), but the aesthetics of (*a*) are greater because the black lines show a greater response to the push of the circle.

(*a*)

(*b*)

Fig. 2.13. The crag appears to push the plough lines to the right in a similar way to the abstract in Fig. 2.12(*a*). This looks better than if the plough lines ran straight down the slope.

rectangular element is perceived as having strong directional properties.

Forest shapes also give rise to visual forces, such as the rectangle which creates four diagonal arrow heads. If we consider a diverse range of shapes, their visual forces can be analysed as shown below. When these shapes are combined in a design, the visual forces produce *tension* which is defined as the interaction of visual forces. If the direction of visual forces conflicts, the high level of unresolved tension can produce a disturbing effect. This can be reduced to some degree by repositioning the shapes, but would be further improved by the selection of more compatible shapes. A small degree of tension can add interest to a design.

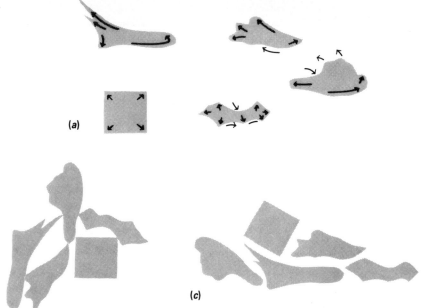

Fig. 2.14. (*a*) Visual forces in five shapes. (*b*) The shapes combined, with a high level of unresolved tension. (*c*) A different arrangement of shapes reduces the tension, producing a more unified design.

The key point is that highly disruptive and intrusive effects occur where forest shapes conflict with visual forces of the landscape.

Visual forces in landscape

There are consistent visual forces in all but the very flattest landscapes. The eye is drawn along a receding sinuous road, and the contrast of sky and water with landform also draws the attention. Skylines and shorelines have a major effect on any composition.

Fig. 2.15. (a) Llyn Brianne, Dyfed. (b) The eye is drawn, bouncing, from one spur to another. (c) The spurs appear drawn together, each into the bay opposite. (d) The eye is drawn downwards on the spurs.

(a)

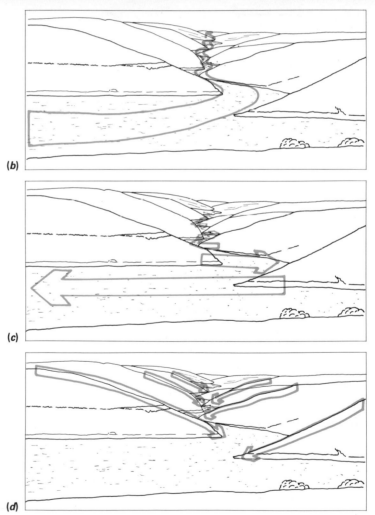

(b)

(c)

(d)

(a)

Fig. 2.16. (*a*) Analysis of visual forces in a typical upland landscape. The strongest arrows illustrate the largest and most pronounced forms. Smaller arrows indicate more subtle prominences and depressions on the slopes. (*b*) The shapes of forest margins responding to visual forces in the landform creating a direct and satisfactory relationship between the forest cover and the landscape. (*c*) A similar forest shape would be just as visually satisfying on the upper slope, though lower slopes should be planted at the sides to provide better balance. In practice, the rocks would not be planted.

(b)

(c)

The example of Llyn Brianne illustrates a general tendency for visual forces to draw the eye downwards on spurs, and up into valleys and concavities. Observation of landform in a wide range of landscapes has shown this to be so and that a hierarchy of visual forces exists. It is, therefore, possible to analyse a typical upland landscape in terms of visual forces related to the convexity or concavity of the landform.

Given the need for forest shapes to relate to landform, it follows that an edge running across a slope should be designed to rise and fall in response to the visual forces. This applies whether trees are to be above or below the line.

The relationship of visual forces to landform can be plotted with accuracy on a contour map, but because of the great range of other factors which come into play, slavish following of a detailed analysis is unlikely to produce a good design. Careful study of visual forces can, however, give useful cues to design of forest shapes. The principle can be applied to any lines in the landscape, at any scale and on dramatic or subtle landform.

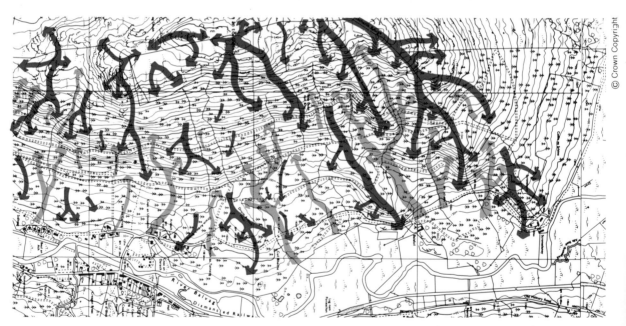

© Crown Copyright

Fig. 2.17. Visual forces in landform shown on a contour map.

3. Scale

- Scale is concerned with relative size and has a major effect on our perception.

- The scale of woodland and forest should reflect the scale of the landscape.

- The scale of the landscape increases the further you can see, the wider the view, and the greater the difference in elevation.

- Small scale irregular shapes may appear geometric at a greater distance.

- Areas appear to be different sizes and shapes when seen from different viewpoints.

- The scale of the landscape is greater on higher slopes and hill tops than on lower slopes and in valleys.

- The scale of landscape design should vary gradually from one area to another.

- Where the landscape is divided into different parts, a ratio of one-third to two-thirds is generally most satisfying, e.g. a hillside one-third open and two-thirds wooded.

Scale is the comparison of size of one visual element to another, and to the size of the whole composition or to ourselves. The relationship of size to ourselves can be illustrated by comparing various sizes of stones with the human figure, the latter providing a reference to which other sizes can be compared.

Fig. 2.18. Pebbles are seen as fine texture. Small rocks are seen as coarse texture with individual forms also apparent. Boulders are seen as forms: close to a large boulder the surface texture dominates, while at greater distance its form is re-asserted.

Scale of the landscape

The scale of a landscape seen from any point depends on how much of it is visible in terms of the three dimensions, i.e. the vertical height of a hill, the breadth of the view, and its distance away. If we can see a great distance or a wide panorama, the scale of the landscape is large. It is also large if one can see great differences in elevation.

(a)

(b)

(c)

Fig. 2.19. (a) Crosscliffe Viewpoint, North Yorkshire Moors. A large scale landscape of broad and distant horizons. Any woodland on the scarp or the moor should be in large sweeps (b) rather than small fragmented areas (c).

In valleys where the views are more contained the scale of the landscape tends to be medium to small scale.

On flat ground trees and forest edges often define the scale of the landscape.

Fig. 2.20. A medium to small scale valley landscape in the Lake District.

Fig. 2.21. A small scale landscape in Strathdee, Aberdeenshire.

Fig. 2.22. Trees and forest edges defining the scale of the landscape. If the edge is too far away, control of the scale may be lost.

Fig. 2.23. Where the forest is seen at close quarters from picnic places, footpaths, or roadsides the scale of the landscape is small, and details are more obvious and important.

Scale and viewpoint

Scale also varies with the viewpoint of the observer. In hilly terrain the scale and shape of an area seen from below appears small, but increases as the viewpoint moves higher. Similar effects occur when an area is seen from different directions in a horizontal plane. In narrow valleys, where oblique views often dominate, this effect can appear disruptive. It may be overcome by screening some areas substantially or by extending their shapes sufficiently along the contours to increase apparent size.

Occasionally, elements on opposite sides of a narrow valley, e.g. felling coupes, appear to merge creating an intrusive large scale. They may have looked all right when designed from opposite sides of the valley, but the problem only shows up in views along the valley.

Scale and distance

Viewing distance greatly affects the interaction of scale and shape. A forest edge seen from a few hundred metres may appear irregular, but shows up as a straight line from a greater distance. Small scale irregularity is best accommodated on broader shapes already designed to conform to the larger scale of the more distant viewpoints; if this is done well it looks right at any scale.

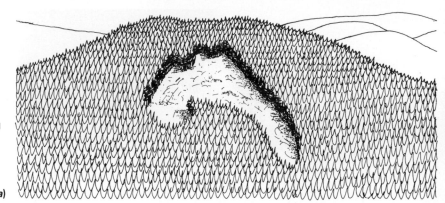

Fig. 2.24. As an area is seen at a more oblique angle (*a*) it appears smaller (*b*) and eventually disappears (*c*).

(*a*)

(*b*)

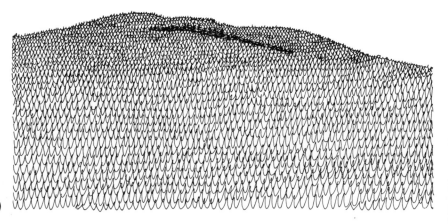

(*c*)

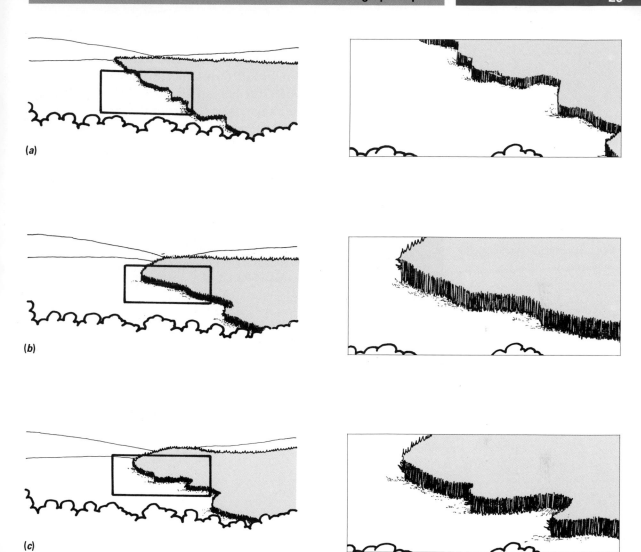

Fig. 2.25. (a) An irregular edge can appear straight from a greater distance. (b) What looks like a good broad curve from a distance may seem boring in the shorter view. (c) Irregular detail overlaid on broad shapes looks right in short and long views.

Scale and topography

Topography determines the extent of views and, therefore, the scale of the landscape in broad terms. Scale increases as views extend and is, thus, greater further uphill. It is the task of landscape design to reflect the range of scales, from the largest near the skylines to the smallest in valleys and places which are intensively used, e.g. roads, villages, car parks, etc.

(a)

(b)

Fig. 2.26. (a) Large scale landscape near the skyline with smaller scale in the valley bottom in Ennerdale, Cumbria (courtesy Mike Pearson). (b) The change of scale from one part of the landscape to another should take place gradually and abrupt changes be avoided.

Scale and proportion

When the scale of an element appears too big it can be reduced by subdividing into different parts. The scale of a building can be reduced by using different materials, which also adds interest.

Fig. 2.27. Effective use of colour breaks up what would otherwise be massive scale of terraced houses at Jedburgh, Roxburghshire.

When elements are divided in this way asymmetric proportions of one-third to two-thirds are generally the most satisfactory. Further subdivision can be done in the same proportion. A half and half division rarely looks right, usually producing an uneasy symmetry, particularly in the broader landscape.

In a landscape composition the 1:2 proportion is useful in attaining a satisfactory balance between elements in a particular view, such as between open space and woodland, or between broadleaves and conifers.

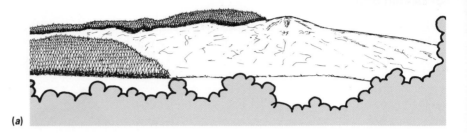

(a)

(b)

(c)

Fig. 2.28. (a) Despite a clear dominance of open space at 73 per cent, the scale of the upper remnant of forest is too small and the composition is unbalanced. (b) Proportions on a roughly one-third to two-thirds basis produces a far more comfortable scale. (c) Nearly equal amounts of open space and forest produce a far less resolved result.

4. Diversity

• There is a basic need for visual diversity in man's environment.

• As a result of varied geology, climate, and human use, the British landscape is highly diverse.

• Excessive diversity in a landscape or design often leads to visual disruption and confusion.

• Ecological diversity can contribute to landscape diversity, but the two are not necessarily equivalent.

• Afforestation can reduce landscape diversity if it hides landscape detail beneath the tree canopy.

Diversity is the degree and number of differences in a landscape or design.

Fig. 2.29. Diverse landscape of Strathtay, Perthshire.

The need for diversity in the environment has been recognized by successive generations of landscape designers and, more recently, by psychologists. The latter generally associate it with quality of life and emotional well-being. J. C. Loudon (1833) quotes Humphry Repton (1752–1818) as identifying variety and intricacy as desirable attributes in landscape designs. Two centuries later, the value of complexity and variety continues to be noted by landscape architects, e.g. Litton (1972) and described by psychologists such as Kaplan (1973) in a similar way. This basic need may have arisen from primitive man's perception of a diverse landscape as offering better food supplies and opportunities for shelter, defence, and chance to see predators, competitors, or prey.

Landscape diversity in Britain

Varied climate and geology, and the long history of human occupation contribute to the very diverse range of landscapes present in Britain, a relatively small country. Man has left many marks on the landscape, from prehistoric earthworks and megaliths to medieval field patterns, buildings of various ages, and designed landscapes. Archaeological value, or historic or literary associations make some landscapes candidates for conservation, though this does not mean that they should invariably be preserved intact at a particular stage of their development or history. The features, qualities, and character which are of particular significance should influence the planning of any landscape change. If the diversity of our national landscape heritage is to be maintained, it is important not to destroy or obscure special landscapes, and to recognize and respond to the spirit of the individual place.

The level of diversity in design

The requirements of diversity have to be balanced by the need for an overall effect in which all the elements contribute to the unity of the whole landscape composition. Increasing diversity has a number of consequences. In a monotonous composition it can add interest, but more elements will interact and more organization of the design is necessary if visual conflict is to be avoided.

Diversity and scale

Scale tends to be reduced when diversity is increased, and where scale is too large this is often an improvement. Where scale is too small or fussy there has to be a greater degree of organization. If a broader pattern cannot be established, a reduction in overall diversity may be required. Scale is important in this respect because a high level of diversity can be sustained if one element is clearly dominant. Humphrey Repton considered that all plantings should be dominated by one species occupying two-thirds of the whole; the remaining one-third could then consist of a wide variety of

Fig. 2.30. The introduction of additional diversity may lead to visual disruption (courtesy RSPB).

Fig. 2.31. A similar problem of numerous signs and structures, better resolved by repeating similar size, shape, and colour.

plants. Other devices such as the grouping of similar elements and the development of structure can be used to prevent conflict between diversity and unity. This can be a particular challenge, in that some elements such as crags are already in position and cannot be moved.

Excessive diversity in the design of small buildings and structures can make them appear even smaller and at odds with natural landscape qualities; it should, therefore, be avoided.

Ecological diversity and landscape diversity

Variation in stand structure, the presence of open habitats, and a variety of tree, shrub, and ground vegetation species can often contribute to both ecological and landscape diversity. The two do not always equate, however. If ecological diversity is too intimately mixed it may be seen as a uniformly confusing tangle; an overview of an ecologically rich natural forest may appear as nothing more than a large mass of dark green texture. Diverse landscapes, on the other hand, may be ecologically uninteresting.

Fig. 2.32. Beechwoods in the Chilterns, Buckinghamshire. A visually diverse forest landscape of limited ecological diversity.

Fig. 2.33. The ancient and ornamental woodland of the New Forest, Hampshire, where ecological and landscape diversity occur together (courtesy Terry Heathcote).

A landscape composed of one tree, one shrub, and one grass species may exhibit diversity derived from visual variables, yet be ecologically impoverished. Many urban landscapes have considerable diversity and very low ecological value.

One of the most demanding objectives of forest landscape design is to develop appropriate landscape and ecological diversity so that either can be emphasized in different circumstances.

Effects of afforestation on landscape diversity

Afforestation can reduce landscape diversity by hiding features and detail beneath the trees, although the introduction of woodland elements into some landscapes may increase diversity. Variety of species and ages within the forest adds interest, but may be an inadequate substitute for significant loss of unusual or characteristic features. Stand diversity should be developed to balance the disappearance of the ordinary detail of boulders, vegetation patterns, and small streams beneath the tree canopy. The reduction, maintenance, or increase of landscape diversity is one of the most important and easiest factors to assess when deciding whether a proposed change is detrimental or an improvement to the landscape.

Fig. 2.34. Pattern of rocks, grass, and heather obscured by planted forest (courtesy George Dey).

5. Unity

- Unity is an essential objective in landscape design.

- Unity is achieved when visual contrasts in a composition appear to be counterbalanced by visual similarities and the various parts are organized into a clearly identifiable composition.

- Variable factors such as shape, visual force, and scale should be used to unite the forest with the landscape and counterbalance contrasts of colour, texture, and form.

- Landform has a major influence on landscape character and forest shape should be related to it in order to develop unity with the landscape.

- Interlocking shapes can be used to increase visual unity.

Diversity in landscape provides interest and contrasts without which everything becomes background; but if the number and degree of contrasts are too great, unity is lost in confusion, and discord. The balance between contrast and continuity can be illustrated in the following examples.

The simplest is an abstract of a black L-shaped motif repeated on a white background, as in Design A in Fig. 2.35. Although the L-shape provides a theme, the way it is used is highly varied in terms of its size, proportions, and number. The collection of small L-shapes appears as a textured rectangle and because the thinnest L runs beyond the frame, it appears to be the corner of a larger shape. The form of the L motif is so varied that differences in position and direction create a disjointed overall effect.

In Design B the L-shapes are laid out parallel and at right angles, and are more closely grouped. The various elements are organized to re-create the L motif at a more appropriate scale by means of defined enclosures of space ('closure'). A unified design results because the level of contrast is balanced by greater continuity of scale and direction. This balance of contrast and similarity is a main objective of design, as important in landscape as in other visual arts.

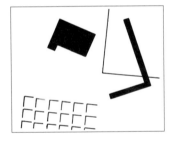

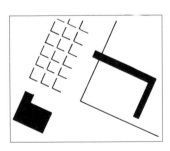

Fig 2.35. Abstract development of unity.

Unity in the landscape

Design in the abstract example above is simple because the background is a blank sheet of paper. Natural patterns and artificial

Fig. 2.36. A small structure completely disrupted by too much diversity. Eight different finishes and three separate forms would have been appropriate only on a much larger building.

features already present in the landscape affect the appearance of any new element.

In the example below the large light-coloured shed stands out from the surroundings in marked contrast to the other buildings. Its contrast with the landscape in terms of scale, shape, colour, and texture is insufficiently balanced by any factor uniting it with the landscape, and it looks intrusive as a result. Without the shed, the buildings show sufficient continuity with the background to appear as an interesting and inevitable part of the scene, and still be sufficiently different to contribute to diversity. It is vital to have a balance of sufficient contrast to maintain diversity and enough similarity to avoid intrusion.

In comparison with buildings, the impact of woodlands and forests is on a far larger scale. They must be designed to blend with the landscape, as background rather than as contrasting elements.

Fig. 2.37. Large farm building in the landscape.

We can identify factors which inevitably cause contrast and others which can be manipulated to achieve unity with the landscape. Even-aged conifer forests tend to be highly unified within themselves, but certain factors create unavoidable contrast with open ground.

- Much darker colour and increased shadows, particularly with evergreens.

- Coarser texture, as a result of cultivation and later from shape of tree crowns.

- Increase in vertical height as trees grow taller.

Other factors can create contrast with the landscape, but can be adjusted to help to overcome the inherent contrasts above.

- Shapes of external margins, species, open space, etc.

- Loss of diversity by screening surface detail by trees.

- Increase in scale of small-scale landscapes and a tendency to reduce scale of large-scale landscapes.

- Abrupt vertical edges, particularly of conifer woods.

When felling and restocking take place, all the contrasts with the surrounding landscape are present, and the additional elements of felling coupes and varied ages are introduced. The

Fig. 2.38. Forest unified with landscape at Achray Forest, Perthshire. The contrasts of colour and texture of evergreen conifers with the surrounding landscape is counterbalanced by: diversity introduced by deciduous larch; similar shapes and colour of larch and bracken; forest edge and evergreen spruce running up the depressions left and centre, in good relation ship to landform.

increase in diversity, the shapes of differently aged crops, the apparent scale of the forest, and so on can all be manipuated to enhance visual unity within the forest and with its surrounding landscape.

Unity and shapes

The design of forest shapes in conformity with landform is a powerful means of developing unity with the landscape. Unity can be increased by designing boundaries between shapes so that they penetrate and interlock with one another. This as an important device in uniting forest and open ground.

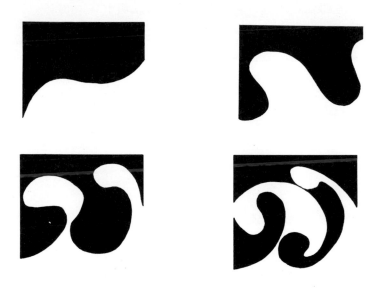

Fig. 2.39. Increased interlocking of one shape with another increases unity, but if carried too far can lead to fragmentation. Similar effects are seen in landscape and this is an important device for blending forest shapes with open space.

Fig. 2.40. Unity of forest, open ground, and landform developed by interlocking shapes at Tighnabruaich Forest, Argyll.

Fig. 2.41. Felling coupe which increases the interlock of open space, landform spur, and forest shapes at Ardgartan Forest, Argyll.

Fig. 2.42. Interlock of open space and small broadleaved woodlands, increasing unity of landscape (Chilterns, Buckinghamshire).

6. Spirit of place

- The spirit of each individual place or **genius loci** is unique and should be conserved and enhanced in forest landscapes. It is most fragile where it is most obviously present.

- *Genius loci* is important as an asset, a stimulus, and an objective of design.

- Although individual features are sometimes associated with *genius loci* it is more often expressed by unique contrasts or combinations of such features.

- Features contributing to the spirit of a place should not be hidden unnecessarily, but should be emphasized by forest design.

The spirit of place or *genius loci* is the quality or qualities in a landscape which make it unique and special.

It is often appreciated on a subconscious, emotional level and the idea is probably easier to grasp by considering illustrated examples, by looking at paintings and sculptures, and by reading literary descriptions of familiar landscapes. A comprehensive analysis of the subject is given by C. Norberg-Schulz (1980) in his book entitled *Genius loci*.

Norberg-Schulz points out the importance of place in our lives. We may identify ourselves and each other by where we come from, as in 'I am a Londoner'. Place is more than a location: it is the total effect of a set of natural and man-made objects, located in a unique way. The effect of space and character is important to the sense of place. By space, we mean whether a place is open or closed, above or below its surroundings, and so on. All places have character, usually described by adjectives such as barren, fertile, harsh, gentle, etc. Norberg-Schulz explains the difference between character and identity, in that the latter implies an individuality which we can easily understand. This individuality is an important part of *genius loci*.

Certain characteristics are often associated with a powerful spirit of place; strongly defined space, or very exposed summits and high points, for example. Indications of great age and maturity such as old trees and rock often engender *genius loci* especially when associated with water. It often occurs where the essence of a wider landscape, in terms of landform, trees, and enclosure, are distilled into a smaller space. Ambiguity and appearance of wildness can also contribute to the phenomenon.

Reference to writers' and artists' portrayals of a landscape helps to identify *genius loci* and is a useful preliminary to landscape design. Much of the British landscape has been recorded by painters who have often succeeded in capturing the essence of the place. The opening passage of *The return of the native* by Thomas Hardy is an excellent example of the expression of *genius loci*.

Genius loci and landscape design

Although we can create well-unified, diverse forest landscapes, they may appear bland and uninteresting unless each place has its own individual character. Great care is needed wherever the spirit of place is most obvious; it is an elusive quality, easier to conserve than to create. Peoples' emotional attachment to the landscape is strongest where there is a strong *genius loci*, and they are likely

Fig. 2.43. Glencoe, Argyll

to object to landscape changes in such places. The features and qualities which contribute to it should be identified, conserved, and enhanced.

Spirit of place can be a stimulus to design and help to identify design objectives for individual sites. In a dramatic landscape such as Glencoe, *genius loci* is strong. It is expressed in the brooding magnificence arising from the strongly enclosing and massive vertical landform. The scale is vast and the darkness of the rocks increased by the deep shadows present for much of the year on the southern side.

From the floor of the glen massive boulders dwarf the visitor and heighten the impression of enormous scale. The absence of trees means that the mountain walls are always in view. Whichever way one travels through it, the space is gradually constrained until the most impressive part of the pass is reached. Here, the strongly enclosed space and the abrasive qualities of the rocks make the most powerful impression. Weather conditions of cloud and dramatic lighting add to the spirit of place, intensified by recollections of the massacre with which Glencoe, the 'Glen of weeping', is linked.

Such effects should not be set aside lightly. This is not a place for gaiety and light-hearted design which could not compete with the atmosphere of the place. There is little scope for woodland, but some planting with sombre evergreens, particularly indigenous Scots pine, would be in keeping with the *genius loci*; the more

Fig. 2.44. Corrieshalloch Falls, Wester Ross

Fig. 2.45. Forest character creating *genius loci*.

strident colours of Japanese larch, for example, would appear garish and incongruous.

More intimate places also develop a powerful identity. At Corrieshalloch in Wester Ross the waterfall is spectacular in itself, and the context of the ravine with its crags, undisturbed vegetation, and mist drifting off the fall gives rise to a strong spirit of place. The effect of the gorge plunging to earth far below the viewpoint perched on its edge is outstanding, and the appearance of seclusion and maturity contribute to the spirit of place in quite a different way from Glencoe.

Genius loci in conifer forests often derives from the great size of trees and their soaring vertical trunks. The particular quality of broadleaved trees is expressed more in their diversity of form, particularly in ancient, open-grown or pollarded specimens. Signs of great age and decay also contribute.

Fig. 2.46. Natural landscape quality is compromised by urban white lines and painted barriers in the car park from which this view is seen. An unmarked car park divided into bays would have respected the wild quality of the place; so would a darker colour of markings and barriers, reflecting colours of the surrounding landscape.

Conservation of genius loci

Introduction of even the smallest structures can impair *genius loci*. Facilities for visitors are important in this respect as they provide the context from which the place is seen. The spirit of place which attracts visitors can be needlessly spoilt by poor design of the structures needed to give access to the site. The first artificial element introduced into a wild landscape is always highly significant, and should be designed for minimum impact and a respect for natural qualities.

Landform with exceptional sculptural qualities can be affected by insensitive afforestation, but it is often the case that some parts of a landscape have less significance, and with careful design some planting can take place without detriment to the *genius loci*.

Fig. 2.47. In the Brecon Beacons National Park the landform of curving scarps has sculptural qualities which are brought out when lit from the side. To preserve the *genius loci* planting should be kept well clear of the ridges and scarps.

Genius loci as a stimulus for design

The spirit of place should influence the design of buildings or other artefacts in a particular locality. A direct relationship can be created between a building and its site or history. An example of this is the Visitor Centre at the Sherwood Forest Country Park, where the buildings were designed to echo the form of the charcoal kilns which were once a feature of the area.

Fig. 2.48. The Visitor Centre at Sherwood Forest, Nottinghamshire. The form of the building reflects that of the charcoal burners kilns which were once a feature of the forest.

In conclusion, the six design principles discussed in this chapter are cardinal to good forest design. Their application is discussed at greater detail in later chapters. With an understanding of these principles, *the landscape provides the guidance for design*, if we take the time to look at it carefully.

Concepts of forest design

- As natural an appearance as possible should be sought and urban layouts, materials, and structures avoided.

- Forest shapes should follow landform.

- The forest should reflect the scale of the landscape.

- Within the forest, visual and habitat diversity should be developed as much as possible.

- The forest should blend with the landscape.

Conservation of natural landscape

In addition to the general principles described in the last chapter it is useful to identify broad aims for forest design which will conserve and enhance the landscape. Man's use has such a consistently urbanizing or domesticating effect on its appearance that the diversity of the more natural landscapes may become lost. This could eventually deprive our urban population of the relief that so many crave from their artificial daily environment and reduce the variety of landscapes to be seen in Britain.

It is useful to consider, first of all, how natural and urban landscapes differ. They have highly contrasting and complementary characteristics, and as land use intensifies, the natural character of the countryside becomes submerged by more humanized patterns unless a natural appearance is deliberately planned. Although the visual characteristics of both are equally valid in their place, natural character should be conserved in the forest wherever possible.

In earlier times the forest was regarded as wilderness, limitless in extent, in which there were clearings which contained settlements. In contrast, many areas of recently established forest are set against a background of open land, created by centuries of deforestation. It seems illogical that geometric pattern is accepted in agricultural landscapes, but not in man-made forests. The reason

Fig. 3.1. A landscape in Snowdonia where, despite man's influence on the vegetation and land use, the overwhelming impression is of wildness. Natural forms and patterns, and the scale of the landscape dominate man's presence, confined to the valley bottom.

is that the forest increases both scale and visual impact. The slight natural variations of landform which remain visible under agriculture become completely dominated by the mass of a geometric woodland shape. While straight hedgerows have similar crops on either side, the edge of the forest is strongly emphasized by changes in height, introducing shadows, and increased scale. There is also a stronger and lasting contrast of texture and colour between the forest and open ground. The pattern of agricultural enclosures varies much more, especially when compared with the rectangular grid of rides present in many forests.

Because of these differences, the geometry of agriculture distracts less from the natural aspects of the landscape than that of forestry. Forests are often on steeper ground, where intrusive appearance is more noticeabe. These areas tend to have a more natural pattern of grass, bracken, rock, and scree, because this is the land which is farmed less intensively and can be spared from agriculture.

The characteristic sheer extent of the forest is given emphasis by the free curves of natural landform. Straight lines placed extensively on the landscape cause unresolved conflict with natural influences, and degrade the interesting contrasts provided by the smaller geometric forms of farms and settlements. Conversely, natural patterns and shapes in the forest will emphasize both the natural qualities of landform and its extensive nature, creating an attractive contrast with smaller buildings and settlements.

Landform and forest shapes

The design of the forest can affect the appearance of the landform, although the latter usually dominates. The pattern of individual trees, as in hedgerows, only becomes prominent where the land-

(a)

(b)

Fig. 3.2. The acceptable geometry of agriculture appears too unnatural when translated to the larger scale, greater contrast, or steeper slopes of the forest.

form is gentler, notably in the lowlands. If the pattern is weakened the landform readily regains its dominance. This may occur where trees have been felled, or where open land is planted and appears continuous with the hedgerows. The broad landform provides the underlying form of the forest, and an unchanging background; the forest must be designed to conform to it.

Reflecting the scale of the landscape

Having a major effect on perception of our surroundings, scale greatly affects the degree to which the forest blends with the landscape. In all but the flattest topography, landform defines the general scale of the landscape, i.e. how far one can see. Larger variations in the forest landscape are needed in longer views, while smaller details are more appropriate in places seen from a short distance.

It is a mistake to imagine that the visual impact of woodland can be reduced by breaking it into smaller parts. This creates a very 'busy' appearance which conflicts with the natural continuity of landform. On the other hand, a large scale pattern of forest in a smaller landscape appears monotonous and brutal. A balance is necessary which reflects the varying scale of the landscape from one part to another. This means that some large areas of continuous forest are needed to form a background to the more diverse patterns of open ground vegetation, water, and settlement in the middle or foreground.

Design of buildings and structures

The natural qualities of the forest landscape should appear to dominate any buildings and structures. Urban and suburban designs for offices, workshops, cabins, fences, gates, car parks, etc., should be avoided in a forest setting. Where there is no rural equivalent for a particular structure, it should be designed simply and anonymously in scale with the landscape in order to minimize distraction from the natural background. A simple, single form provides sufficient contrast to add interest to the landscape; as the number of structures increases they can become very confusing. Where a number of structures is required, as in a car park, they should be kept to a minimum, be as similar as possible, and preferably with dark and subdued colours.

(a)

(b)

Fig. 3.3. The set kerbs, white lines, and *Hypericum* in the car park above (a) have strong suburban associations. In the example below, (b), the surface of local stone and containment by trees integrate the car park strongly with the landscape.

The broad landscape

The broader aspects of the landscape often define the essence of a place. So that the local qualities of the landscape are not lost, we have to consider sensitivity, heritage, context, character, and space before involving ourselves in the landscape details.

Landscape change

Some changes in land use conflict with the qualities of important landscapes to such an extent that they should not be carried out unless the difficulties can be overcome by good design.

Wherever possible, extensive and radical landscape changes such as afforestation should enhance the landscape and counteract people's dislike of change. This emotional reaction is a natural result of feelings of insecurity and anxiety, as familiar surroundings are changed by unknown influences. There may also be a preference for open landscapes, notably by walkers. The landform and vegetation patterns of hill and moor are often perceived as highly natural and any major landscape change is unwelcome, be it afforestation, the construction of reservoirs, or whatever. In contrast to the more truly 'natural' pattern of earlier times, in which artificial clearings were made in the forest, the open hills now seem to be the natural background against which woodland and forests appear as a man-made element.

This difference between what *is* natural and what *looks* natural has bedevilled discussion about landscape change. The distinction is important, because opportunities to develop natural looking shapes have often been missed while debate becomes polarized between the merits of 'natural' oak and 'alien' spruce. The artificial appearance of a square plantation of even-aged trees looks out of place in the wider landscape, regardless of the species planted.

In the longer term, public reaction to landscape change may be diminished by blending changes into the landscape with care, and by maintaining a proportion of open space, especially near walking routes. Good design standards have to be widely applied and continually improved to keep pace with rising public expectations.

Fig. 4.1. An appearance of wildness and the presence of water enhance the quality of the landscape.

Landscape sensitivity

- The sensitivity of a landscape is affected by its intrinsic quality, its visibility, and recreational use.

- The highest landscape quality is defined nationally by statutory designation.

- The quality of individual sites can be assessed in terms of unity, diversity, spirit of place, apparent naturalness, and absence of eyesores.

- The visibility of an area is affected by the number of people who see it, its elevation, and steepness.

- Permanent residents increase the sensitivity of a landscape because they see it for longer periods.

- Enjoyment of the landscape is a major reason for visiting the countryside, so much-visited areas are specially sensitive.

Landscape sensitivity derives from judgements of landscape quality, how visible it is to the public, and the degree of recreational use.

Landscape quality

The quality of individual landscapes is assessed by observation, experience, and judgement. It is very difficult to compare the qualities of two landscapes of different character. The best areas of the national landscape have been identified as National Parks, Heritage Coasts, and Areas of Outstanding Natural Beauty (AONB) in England and Wales, and as National Scenic Areas and Regional Parks in Scotland.

The conservation of natural beauty is a major objective in all these areas. The National Parks are also intended to provide for public recreation. All of them require high standards of landscape design.

The National Scenic Areas in Scotland are smaller, more numerous, and more defined in landscape terms than the National Parks or AONBs. Their locations, extent, and descriptions are published in *Scotland's Scenic Heritage* (Countryside Commission for Scotland 1978). This is a useful reference which describes individually the particular features of this wide range of landscapes.

There is greater variation within the more extensive National Parks of England and Wales, in both landscape quality and character. Some have outstanding 'core' areas which are extremely sensitive, and which may be identified in National Park plans. The parks all contain areas of generally attractive quality which form a background to outstanding places with much stronger *genius loci*.

The designated areas provide a high base level by which to judge individual landscapes, but in all cases quality must be individually assessed. Intuition, judgement, and experience should all be applied to this task in an open-minded way, and diversity, unity, and spirit of place evaluated as guides to quality. Wild appearance, presence of water, and absence of eyesores are additional factors which raise the quality of the landscape.

Visibility

The visibility of an area is affected by its steepness and position in the landscape, as well as the number of people who may see it. Areas at high elevations can usually be seen from greater distances, while those on steeper ground have more visual impact. These aspects should be assessed by examination of maps and by ground observation.

The number of people viewing the area can be estimated in terms of the general population, number of houses, workplaces, etc., and the status and use of access roads. Opinions of local residents should be given due weight because of the greater time they may spend within view of the area.

The time visitors spend on recreation also increases the level of landscape sensitivity; even though numbers may be low, the enjoyment of that particular part of the countryside is often the main purpose of the visit. Forest-based recreation requires an attractive environment and providing access alone is not enough. Walkers are particularly knowledgeable and aware of their surroundings, and landscapes visible from walking routes require high standards of design and sufficient open space to give an interesting sequence of views.

Landscape heritage

- Landscapes may be valued for their history, traditions, and association with artistic works as well as for their aesthetic quality.

- The most important and least changed landscapes should have priority for conservation.

- A balance should be sought between aesthetic quality and other aspects of the landscape.

- Changes in the distribution of woodland and open space in 'designed landscape' parks should be made with the utmost care.

Certain landscapes are valued as the settings for historic or literary events, for their association with music, or as the subject of paintings. Traditional field patterns or the remnants of early industry are also worthy of conservation. Landscapes which appear least

Fig. 4.2. A stone circle in Fernworthy Forest, Dartmoor. Although the forest is kept well back from the monument, visitors cannot appreciate its original setting nor tell whether it has any significant solar or lunar alignments.

changed from the period or work with which they are associated are likely to be the most valued and also the most vulnerable. The visual effects of woodlands and forest in such locations must be accurately assessed; it may be possible to enhance the impression of the relevant period by planting and appropriate woodland management.

On important historic sites it is open to question as to whether it is possible or desirable to attempt detailed restoration of the supposed conditions of the time, and an impression is often more satisfactory. Major battlefields, for example, are best kept sufficiently open for the events of the day to be understood.

Traditional, spiritual, and astronomical sites

Sites associated with traditional festivals and ceremonies, and those which are believed to have particular spiritual, religious, or mystic significance, are valued to varying degree. Certain archaeological remains, particularly megaliths with marked orientation and linear features, have strong relationships with the landscape. It is, therefore, important that any woodland established in the vicinity of such monuments should not detract from the significance of the site or from the visitors enjoyment.

Designed landscapes and landscaped parks

Woodlands created many of the illusions and enclosures of the landscaped parks of the 18th and 19th centuries. These landscapes represent a contribution to art for which Britain is world famous, and so they should be conserved and sympathetically renewed wherever possible.

In these landscapes the great house provides an important focus which should be maintained if at all possible. Even where the buildings have been allowed to decay, it is often better to maintain a safe ruin than to raze them to the ground and so inevitably destroy the main focus of the park.

The original intentions of the designer, and the condition and status of the landscape should be carefully assessed wherever forestry operations are contemplated. There are many cases where the quality of open spaces has been reduced by a change from pasture to arable, though this can readily be reversed. The felling of woodland or the planting of open ground on the other hand can easily cause great damage and the consequences of any proposed landscape change should be carefully illustrated and evaluated.

The original design of woodland and positioning of clumps was often carried out with great care and subtlety, to frame successive views from house and drives while screening those parts of the wider landscape which reduced the illusion of nature. Peripheral woodlands often defined the scale of the internal parkland landscape; regenerating these woodlands without destroying the sense of enclosure is a major challenge for landscape architects and foresters. Although it may sometimes be difficult to keep the form of the original design, the spirit and quality should be maintained wherever possible.

Associations with artistic works

Landscapes valued for their associations with literature, painting and music are often of great quality. One thinks of Wordsworth in the Lake District, Hardy in Dorset, or Scott in the Trossachs and the Border country. These are general relationships and other landscapes can be identified in a more specific way, for example, the Doone Valley in Exmoor as a setting for *Lorna Doone*. In many cases, the character of the landscape may already have been so changed that conservation is worthwhile only in limited areas.

The extent to which an area should be conserved has to be balanced against the needs of land use. Only a proportion of the whole area of the 'Constable country' of Suffolk could reasonably be expected to maintain its character unchanged indefinitely, whereas the area around Flatford Mill painted by John Constable is more limited and careful conservation is practicable.

Fig. 4.3. The Doone Valley, Exmoor

Industrial heritage

Industries of the past have left their traces on the landscape and their conservation requires a discriminating approach. While some industrial spoil heaps, for example, are unsightly masses which would look much better if re-shaped into more natural forms, others reflect a particular process of tipping or even have a valuable sculptural quality. In some areas, it may be appropriate to reclaim the majority of spoil heaps while retaining one or two as relics. Some structures or buildings should be left undisturbed for the same reason. Where the surrounding landscape has a higher aesthetic value, reclamation may take priority over conservation of the industrial heritage.

Fig. 4.4. Old colliery tips in the Forest of Dean, Gloucestershire. Some tips simply appear derelict while others are interesting landscape features.

Fig. 4.5. Restored slate workings at Ballachulish, Argyll. In the context of a diverse natural landscape, the restoration of aesthetic quality should take priority over the reflection of our industrial past.

Feature and context

- All parts of the landscape can be seen within a general context, but some parts stand out as features.

- Landscape changes should be carefully designed if near a dominant feature so as to maintain and if possible enhance its character.

- Changes can be made more easily in less dominant parts of the landscape, and should blend and not contrast with its natural qualities.

- In more uniform landscapes the forest design should emphasize whatever natural features are present.

- The entrance points of routes into contrasting areas of the landscape should be memorable and attractive places.

Some parts of the landscape appear to stand out as features from others which form a general background or context. For example, a single hill may dominate the floor of a valley, while in the broader context of a featureless plateau, the valley itself may stand out as interesting. There is a hierarchy of national, regional, and local context which affects our perception of every landscape.

Layers of context

Everything has some significance as part of a wider context: the landscape exists between earth and sky, forests are visible in the landscape, clearings are found within forests, and enclosures and houses within clearings.

Identification of a feature is more than a measure of visibility. It involves contrast with the background (especially of landform), *genius loci*, sequence, heritage, and association. All artificial elements in the landscape are symbols of man's activities, needs, and possessions. Fences mean grazing animals, walls mean defence or shelter, roads mean transport. Man-made objects carry so much meaning for us that they stand out from their background even when designed to reflect the landscape.

Natural features also draw our attention by the opportunities they present. The summit of a hill may be seen as a vantage point or place of defence and seen to be more important than the hill face. The unusual also seems important, especially spectacular geological features such as waterfalls and gorges.

Even where the landscape is less outstanding, the overall meanings and relationship of the parts can be understood by identifying the points which draw the eye and stimulate the mind most readily.

Design to enhance the natural features

Landscape changes near features should, therefore, be made with care and with high standards of design, to enhance their presence and contribution to the landscape. The essence of each feature can be more readily understood by naming it, e.g. summit, clearing, grove, scarp, etc., and identifying its position on a map or sketch. Away from the dominant features, landscape changes can often be made without intrusion if they blend with the natural character. It is particularly important that forest design should not introduce contrasts which distract the eye from the natural features of the landscape.

As the number of natural features increases, the significance of each is reduced. Too many elements competing for attention cause visual disruption and loss of unity. To enhance unity and *genius loci* of very diverse landscapes, the forest design should enhance some features and play down others, according to their natural prominence.

In landscapes which appear more uniform, minor features should be highlighted by imaginative design. A view from a road may be improved by framing with trees, or a dip in the skyline might be emphasized by a protrusion in the upper margin of the forest. The spirit of place can often be enhanced by maintaining the natural balance and distribution of features and context which is a unique aspect of many landscapes, while emphasizing the contrast of natural features with their background.

Different design strategies are needed according to whether the landscape context is largely forest or largely open ground. In more open landscapes the contrast of smaller woods tends to compete visually with natural features and woodland should, therefore, be blended strongly with the surroundings. Diversity becomes more important in a largely forested context, and the pattern of different ages and species of trees, and clearings should be designed to emphasize natural features.

Emphasizing boundaries and entrances

The point where routes cross the boundary between one type of landscape and another are remembered as entrances and exits.

Strong landscape character at these points heightens the quality and contrast of the two types. For example, the points where roads enter Forest Parks should have some memorable features nearby, such as strong landform or large trees. Alternatively, the pattern of forest and open space can be planned to create a striking landscape, e.g. by retaining older stands or by planting close to the road.

Case study: Old Man of Storr

The landscape of the Old Man of Storr illustrates how one strong feature can dominate another and how the less striking parts of the landscape can be changed without disrupting the whole.

Storr is on the Isle of Skye. The island has a generally high quality landscape and many historic associations, including those with Prince Charles Edward Stuart. Storr lies 5 miles north of Portree on the Trotternish peninsula, and is seen from the main public road round the island.

The landscape of Trotternish is dominated by a mighty basalt scarp which winds for 30 km to the north of Portree. Its crags are so spectacular that it takes a rock pinnacle with the scale and vertical emphasis of the 'Old Man' to stand out as a feature. Although the 'Old Man' is visible in other views, it is best seen in context with the scarp above the water of Loch Fada. The water edge, landform, the rhythm of the scarp, and the contrast of the

Fig. 4.6. The Old Man of Storr (courtesy Finlay Macrae).

pinnacle with the sky make this a strongly focused landscape of the highest quality. Each successive element of island, peninsula, scarp, pinnacle, and view provides a context for its successor, yet appears as a contrasting feature to its own surroundings.

This outstanding landscape can accommodate woodland on the slope below the scarp if the highest standards of design are adopted. An open 'plinth' of ground is also required round the base of the pinnacle to provide visual support. The upper margin should be designed to enhance the position of the 'Old Man', and to reflect the rhythms and shapes of the scarp to the left.

The left foreground should remain open because it cannot be planted on an adequate scale with the electricity lines present. As a result, any planting on the right would unbalance the composition; the immediate foreground and far shore of Loch Fada should also remain unplanted for an uninterrupted view of the pinnacle.

Provided the highest standards of design are adopted, the less interesting parts of this landscape could be afforested and its

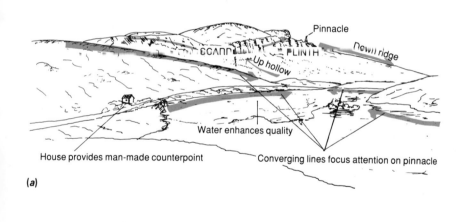

(a)

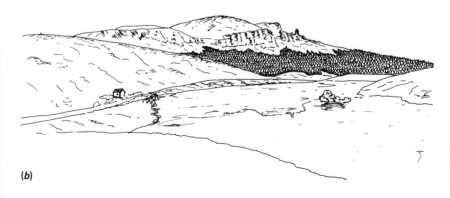

(b)

Fig. 4.7. (a) Naming the most important parts of the landscape helps to identify priorities for design. (b) The less dominant parts of a landscape dominated by strong features can be unobtrusively changed, but design has to be very good.

quality maintained. The features are so dominant that change in other areas appears minimal as long as there is no intrusion by the forest. Landscapes of more subtle quality and character may withstand changes caused by forestry less readily.

Landscape character

- Character is a distinct pattern of elements which occurs consistently in a specific landscape.

- Character can be described in terms of the natural components, human attributes, and aesthetic factors of a landscape.

- Design should enhance the desirable character of the landscape and modify undesirable character.

- Where a desirable landscape character cannot be accommodated in a design the extent to which it is changed should be balanced against its uniqueness and the sensitivity of the landscape.

The character of a landscape can be described as a combination of elements in a distinct and consistent pattern. It is normally used to describe the more widely distributed elements of the landscape rather than those points on which attention is focused. There is no implied value; a landscape may have a desirable or undesirable character. Places may be described as rough textured or open without any implication that one is better than the other. Character does, however, make an important contribution to the spirit of an individual place. By defining the character of a landscape we can assess more readily whether it will be significantly changed by modifications of land use. It also helps to identify whether the range of national or regional landscapes will be reduced by the changes proposed. Character can be used as a guide to design, especially in blending new elements with the landscape; but if too many aspects of the character are emphasized, the essence of the landscape may be lost.

Assessment of landscape character

The landscape can be described in terms of its natural components, human attributes, and aesthetic factors. Landform and vegetation are the most widespread natural components. Human attributes are represented predominantly by patterns of land use, e.g. hedgerows, woods, fields, and buildings. Shape and scale are usually the dominant aesthetic factors. The first step is to identify the combinations of factors which distinguish a particular landscape, such as:

(1) natural components
    landform, e.g low rocky hills
    vegetation, e.g. picturesque old oak, bilberry

(2) human attributes
    field pattern, e.g. walled fields in valley
    settlement and structures, e.g. tight-knit villages,
    scattered stone-built farms

(3) aesthetic factors
    shapes, e.g. square in valleys, irregular on hills,
    diverse green colours

The less important items on the list can then be eliminated and the key qualities of the essential features carefully, but briefly described, in order to define the essence of a particular character. The combination of key elements is often critical; for example, oak and bilberry frequently occur together, but slaty outcrops of rock protruding through an even carpet of bilberry under oak are a very distinctive feature. Potential visual conflicts as a result of forest operations can then be more readily identified. Those which can be avoided by design should be distinguished from those which cannot.

Case study: design to reflect landscape character

Broadleaved trees and woodlands are characteristic of many British lowland landscapes. Their forms have a major impact on skylines and where the upper edge of woodland contrasts with open space. The regeneration of Hang Wood, in Wiltshire, using mixtures of conifers and broadleaves, can be considered against the background of such a landscape.

Parts of the wood were deteriorating scrub and planting was necessary to maintain the range of broadleaves. This operation was only economically viable if a proportion of the area could be replanted in mixture with conifers to provide some early financial return. The conifers would be removed at about 40 years, but until that time they would be increasingly visible through the canopy of broadleaves.

Fig. 4.8. Hang Wood, Wiltshire.

The wood is not in a statutory amenity area or highly exposed to public view, but it has an unspoilt rural quality. It is dominated by the pattern of large hedgerow trees, fields, and small woods, over rounded landform. There is a small, but significant element of conifers in the surrounding landscape, but the character is predominantly broadleaved.

This character can be maintained provided that the proportion of conifers is not too great overall, and that the wood will continue to appear predominantly broadleaved. The latter condition can readily be met by replanting pure broadleaves in areas adjoining the upper edge of the wood. The conifer mixtures should only be planted against a background of broadleaved trees so that their conical crowns are not seen in sharp contrast to the lighter fields beyond. Locating the conifers where they have less visual impact and reducing their overall proportion maintains the particular character of this landscape.

It is not always possible to maintain the character of the landscape by design while changing elements within it. The importance of the character should be balanced with the extent and compatibility of the changes proposed.

The significance of landscape character

The desirability of the character of a landscape must be assessed before deciding whether it should be enhanced, conserved or changed. Even though a diverse range of character ought to be developed nationally, a derelict landscape would only be conserved in exceptional circumstances.

(a)

(b)

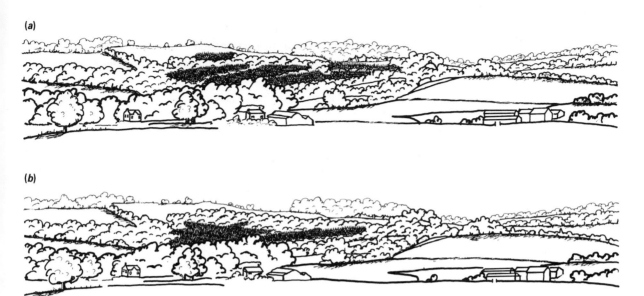

Fig. 4.9. Hang Wood: with the planting of conifers limited to the lower parts of the wood as in (b), the character of the landscape remains predominantly broadleaved.

At the other end of the scale, a particular landscape character might be valued for its aesthetic quality as an important part of a sensitive landscape, or for its traditions or history. Any change in character which can not be alleviated by design must be balanced against the sensitivity of the landscape and its uniqueness. There would, for instance, be a greater need to conserve the best example out of three remnants than a mediocre example of widespread and common landscape character. The expression of the landscape character may also vary sufficiently for it to be conserved where it is most typical and modification allowed where it is least obvious.

The significance of open space and enclosure

- Qualities and variations in space are important aspects of the broader landscape.

- In the rural landscape space is usually defined by landform, hedgerows, or woodland.

- The attractive qualities of space defined by landform should be enhanced, and their unattractive aspects relieved by the design of woodland and forest.

- Substantial open ground may be needed in forested valleys to relieve a sense of oppression or to maintain a characteristic sequence of spaces.

- In forested landscapes open space is important for diversity and should not be planted.

- The overall proportion of woodland to open ground should not be too even and should, ideally, be about one-third to two-thirds, or vice versa, depending on the landscape character.

Space can be defined as an open volume. Its implied proportions and form are recognized as having a deep psychological and emotional effect. The nature and variations of space have a dominant effect on some landscapes. There are infinite variations in the degree and scale of enclosure, from the complete openness of the mountain summit to the confinement of the thicket. Some spaces are significant because of their location, especially at entrance points to contrasting areas of landscape, while others impress with their contrast or proportions.

The influence of landform and woodland

The qualities of space are influenced most by landform in the larger scale, and by the distribution of trees and woodlands in the smaller scale landscape. Although landform is unchanging, the effects of tree growth and forest operations can have considerable influence on the perception of space, particularly forest edges and tree canopy close to the observer. In views across valleys the colour and texture of trees (especially evergreens) increases the impression of enclosure, while in many landscapes the proportion of woodland to open space is very characteristic.

In many landscapes, variation in openness and enclosure can enhance important features as well as the general character, by providing contrast and varied scale. The need for either enclosure or open space depends on situation. Openness is valuable in a specific location such as a hill summit, where it is a component of the *genius loci*: in bland topography where views change slowly the vast extent of sky and sweeping forms is enjoyable for a time, but can become monotonous. Some enclosure in the form of woodland is then desirable because it creates changing views, and greater diversity of form and texture.

The increased sense of enclosure created by forests on valley sides is more desirable in some places than in others. It may augment the spirit of gloomy places with dark colours and poor weather conditions, or it may be an interesting accentuation of the natural constrictions of a valley; but where landform is more uniform, continuous trees along a valley side may transform pleasant and reassuring enclosure into a feeling of oppression. In such cases a proportion of open space should be maintained in the most prominent parts of the landscape to provide some relief.

Sequence of openness and enclosure

The sequence of openness and enclosure along roads and paths should influence the distribution of forest and open ground. It is also an aspect of valley landscapes; the increasing sense of openness as the road ascends from the enclosing woods and hedgerows on the valley floor to the high ground at the head of the valley is a familiar experience. This can easily be lost if the upper slopes are completely forested. A variety of open spaces, especially on the lower side of the road, are often necessary to maintain this traditional sequence.

In some sensitive landscapes the broader pattern of openness and enclosure is such an important characteristic that it dictates the distribution of woodland and open space on a large scale.

Proportions of forest and open space

The proportions of open space to forest as seen in the landscape have a major effect on its character. In mainly forested landscapes open space is an important element of diversity and it is very detrimental to plant up the last remaining open areas. Both the scale of the landscape and that of the individual spaces improve as the apparent proportion of open ground increases towards one-third. Some of the most satisfying landscapes appear to be about two-thirds forest, and carry implications of shelter and productiveness as well as the diversity of texture, ground vegetation, habitat, and open views. Besides providing a background, the forest unifies the landscape at a broader scale and can create, with careful design, a thoroughly natural character.

Landscapes of about one-third woodland overall can be as satisfying as those with one-third open space. The diversity of open views, ground vegetation, rocks, and streams can be seen, and the qualities of landform are readily appreciated, yet there is sufficient shelter and enclosure to provide a setting for habitation, work, and recreation.

The change from two-thirds open ground to two-thirds forest

Fig. 4.10. General factors affecting the landscape at Satran and Crossal.

© Crown Copyright

alters the nature of the background. In a mainly forested landscape the open spaces stand out as features, whereas in the mainly open landscape it is the forest which tends to catch the eye. Either will appear equally satisfactory, unless the other is associated with some rare and highly valued traditional or aesthetic character.

The first forest element introduced into a landscape requires great care. Unless an overall impression of at least one-third woodland can be achieved in perspective, the scale is likely to appear too small and intrusive. Although a small area can be designed to give an impression of larger size, it is better to aim for a visual proportion of at least one-third woodland overall. This is less critical in more intimate or naturally diverse landscapes with strong contrasts, where smaller areas can be more easily introduced onto lower slopes.

Changes as a result of afforestation

Although good design can help to make the landscape changes brought about by afforestation acceptable, there can be cases where proposals are detrimental. This may result in substantial areas having to be left unplanted. Where there is likely to be acute conflict with the landscape, the appearance of the proposed planting should be assessed using accurate sketches or photomontages. The opportunity costs of not planting have to be weighed against the landscape sensitivity, the heritage value, the importance of open space, and the landscape features, character, or sequence that are affected.

Sources of conflict between afforestation and landscape

The following are likely to be the main causes for conflict between afforestation proposals and the requirements of landscape conservation, and which may not be alleviated by design. Good forestry planning and practice will seek to avoid them.

Loss of diversity
1. Planting the last remaining open space in a landscape composition.
2. Screening important views.
3. Completely enclosing long stretches of road with trees.
4. Screening important or characteristic features.

Fig. 4.11. Red Cuillin and Sgurr nan Gillean (see over Leathad na Steiseig, right) from the Dunvegan road (courtesy George Dey).

Landform

1. Change from smooth to coarse forest texture on important ridge lines.
2. Planting of characteristic open skylines to create a predominantly wooded appearance.

Space

1. A sense of oppression created by continuously planted steep valleys.
2. Woodland appears to occupy about half the landscape.

Case study: planting of Satran and Crossal Grazings, Skye

This example shows how the broader aspects of the landscape — sensitivity, character, space, and sequence — influence the distribution of forestry. The area in question lies just outside the Cuillin Hills National Scenic Area in the Isle of Skye, and important views from the main Sligachan–Portree and Sligachan–Dunvegan roads are affected. The quality of the views is very high and the landscape is regarded as extremely sensitive.

These are the only views from public roads in which the contrasting forms of the Red and Black Cuillin can be seen in a single composition. The character of the landscape is clearly expressed in the description of the National Scenic Area (Countryside Commission for Scotland, 1978).

'The jagged gabbro of the Black Cuillin and smooth pink granite of the Red Cuillin combine their contrasting shapes to form a mountain area of dramatic and distinctive outlines of great scenic splendour.... Closer at hand Sligachan is famous for its view of the shapely and serrated Sgurr nan Gillean in marked contrast to the pudding profiles of the Red Cuillin to the east.'

The view from the Portree road is critical to the sequence of landscapes seen on the road from Staffin in the north to Kyleakin, the main ferry terminal. Travelling south towards Portree, the successive glimpse views of the Red Cuillin and the Black Cuillin individually are a foretaste of their combined effect. Near Portree the landscape is more enclosed and this character is maintained on the ascent of the forested Glen Varragill. The openness at the head of the glen creates a sense of anticipation in the latter part of this long, confind climb which is rewarded with this magnificent view of the Red and Black Cuillins. There is time for the motorist to enjoy it before the composition is lost where the descent to

┌─ STRONG CONTRAST ─┐

Simple forms, smooth, even slopes.   Complex forms, concave shapes, jagged, sharp,
              light colours   dull colours

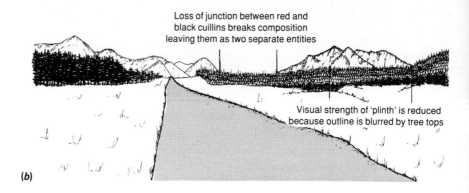

RED CUILLINS   Lines arrowed unite red and black   BLACK CUILLINS
                cuillins into a single composition

Common base uniting Red
and Black Cuillins requires
protection against screening   POSSIBLE ACQUISITION
or blurring of outline

(a)

Loss of junction between red and
black cuillins breaks composition
leaving them as two separate entities

Visual strength of 'plinth' is reduced
because outline is blurred by tree tops

(b)

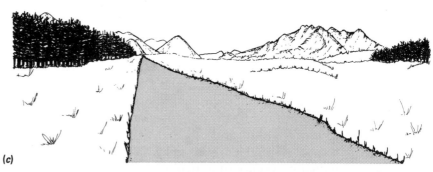

(c)

Fig. 4.12. (a) Landscape appraisal, from the Portree road. (b) Planting across the foreground of the view disrupts the overall composition. (c) Planting on the left side of the view improves balance and frames the composition, but the view to Glamaig (left) will be increasingly obscured as the existing plantation grows. (d) The complete composition of this important view should be revealed by clearance of the existing plantation when timber reaches a marketable size.

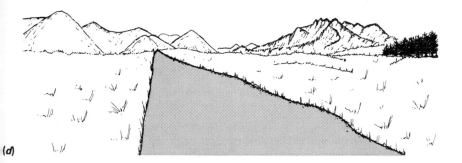

(d)

Sligachan begins. Thereafter, the road runs between the contrasting mountains and they can no longer be taken in at a single glance.

The sequence on the Dunvegan road is less dramatic because there is little anticipation and the view is more gradually revealed. The view is of slightly lower quality because the mountains are closer and Sgurr nan Gillean is seen obliquely over Leathad na Steiseig.

The contrasting shapes of the mountains are composed in the view by the foreground of landform whose smooth texture maintains continuity with the hills. A forested foreground would destroy this continuity. Even though the mountains would be visible beyond a forested edge the key quality of the view would be lost and a substantial area should, therefore, be left open.

Planting can be more readily accommodated in the view from the Dunvegan road, though a high standard of design and a variety of views to individual peaks should be maintained.

Elements of diversity

Understanding the intricate components of the landscape is as important as the view of the broader patterns, discussed in the last chapter. The rich detail of natural and historical features should appear to fit into the broader pattern created by open space, and forest stands of different age and species. A survey recording all the existing features is an important part of the design process. There are so many different details which can be emphasized as part of landscape diversity that scale and unity run the risk of becoming disjointed. Some degree of organization is required, in the form of a permanent framework of features which unify the landscape.

Elements to be considered are:

- *aesthetic features*   open space
                            views

- *topographic features*   water
                            landform
                            rocks, crags, and scree

- *wildlife*   plants
                            animals

- *human influence*   archaeological and historic sites
                            recreation sites, walking routes
                            diverse forest management
                            long-term forest structure

Open space

- Open space within the forest provides important and desirable contrast.

- Open space allows landscape detail to be seen and increases the range of habitats.

- Open spaces are necessary for deer control, protection of water quality, wildlife conservation, picnic sites, and other recreational areas.

- These permanent open spaces are needed in addition to the temporary openings caused by felling and should be part of an overall landscape design.

- Where little open space is required for reasons of timber management in extensive forests, some areas should be left unplanted for landscape diversity and wildlife conservation.

Open space provides an important contrast to woodland and is valuable in its own right. As long as it is appropriate in terms of shape, scale, etc., open space is an asset both inside and at the edge of the forest. It allows other landscape elements to be seen, such as rocks, water, vegetation, views, and landform from viewpoints within the forest.

The contrast of open space

Open space contrasts with the forest in a number of ways. It almost always provides contrast of texture — usually finer, smooth open ground contrasting with the coarseness of tress, though scattered trees and shrubs may give open space much coarser texture than the forest.

There is often additional contrast between lighter colours of grassland and darker trees. The contrast is strengthened by light falling on the flatter surface of the open ground and the shadows present in and below the forest canopy. In areas of darker vegetation such as heather the contrast between forest and open space

Fig. 5.1. Contrast of colour and texture between forest and arable land at Lael Forest, Ross-shire. The scattered broadleaves in the valley provide a third contrast of coarse texture, and the lighter greens of pasture and moorland contrast with the darker forest.

Fig. 5.2. Strong colour contrast between forest and arable land at Alltcailleach Forest, Aberdeenshire. The colour of the forest and the heather moor are so similar that they seem to blend into one in this light. In snow, a stark contrast would appear.

may be less marked, but it is generally more obvious than the difference between tree species.

The sensation of space is important at a deeper emotional level and is a major component of *genius loci*, affecting people travelling through the forest whether on roads or on paths. Varying the openness or enclosure of such routes usually gives a much greater impression of diversity than does planting different species. Maintaining various sizes of open spaces in the forest is, therefore, an objective of design.

Types of open space in the forest landscape

The character and quality of open space varies widely and it is important to distinguish open ground on the woodland edge from that which appears enclosed by the forest. The first is seen as an extension to the broad landscape surrounding the forest, with implications of extent, openness, and diffuseness; the latter provides a welcome contrast and point of interest, with connotations of shelter, light, warmth, and safety. Although open space on the edge is useful in blending the forest with the landscape, it is less likely to have the same sense of place as the enclosed space within.

Open space is particularly important for diversity in extensive forests, whether as agricultural fields, areas of failed crop, areas earlier considered unplantable, etc. Foresters should resist the temptation to plant such ground. Though fellings will open up space within the forest, they are temporary, and a poor substitute for permanent open spaces with mature character and vegetation.

It is necessary to maintain open space in relation to roadsides,

Fig. 5.3. Wareham Forest, Dorset. The contrast of forest and open ground is emphasized by heather in bloom. The enclosed space within the forest has a completely different quality to the extensive areas beyond the edge.

Fig. 5.4. Grizedale Forest Park, Cumbria. The farmland with its pasture and hedgerows contrasts strongly with the forest. Planting the open area would greatly reduce landscape diversity.

Fig. 5.5. Areas of failed crop may represent an economic loss, but in extensive forests small areas such as this are an asset in terms of landscape diversity.

watercourses, rocks and crags, and recreation areas, as well as certain wildlife habitats and glades for deer management. They can often be used for more than one purpose. The benefit to the forest landscape is greatest where they are combined as a linked system of spaces within an overall design.

Open habitat

Specific sites within the forest which are maintained as open ground habitat for wildlife conservation also provide a visual contrast with the woodland. The shape and scale of these open spaces should be taken into account when planning the adjoining forest edge. Where it is desirable to maintain a substantial proportion of

Fig. 5.6. Open ground habitat maintained in the valley bottom adds diversity to the landscape.

Fig. 5.7. Open ground managed for wildlife conservation in the Black Wood of Rannoch, Tummel Forest Park, Perthshire.

open habitat within extensive forest, significant areas may have to be left unplanted, at least until other large open spaces are created by felling. In each case, the precise objectives, essential requirements, and relative importance of wildlife conservation and landscape should be clearly defined as a preliminary to design.

Enclosures

The enclosure of common land and redistribution of land to individual farms which took place in Britain in previous centuries have given a quality to lowland landscapes which is quite different from the more open hills. Certain upland landscapes also show the effects of common enclosures. These can be important elements of landscape diversity and heritage, which should be taken into account in forest design.

Open space as a substitute for species diversity

Loss of landscape diversity resulting from afforestation may be counterbalanced by planting a variety of species, notably larch and broadleaves. These species are difficult or uneconomic to establish on infertile sites, and choice may be limited to one or two evergreen conifers. A greater proportion of open space may be necessary on such areas in order to maintain an appropriate scale and level of diversity.

Open space is, therefore, a most important element of diversity in the forest landscape. It provides strong visual contrast to the wooded area, allows other elements to be seen, and has benefits for nature conservation. The design of open spaces within the forest is discussed in later chapters.

Views

- Extensive views are characteristic of the British landscape.

- Open views make an important contrast to the enclosure of the forest.

- Certain landscapes have compositions of particular quality which should be enhanced by forest design.

- Identification of points from which key views are seen and in which direction should be carried out during landscape appraisal.

(a)

(b)

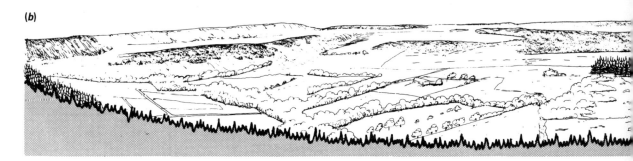

Fig. 5,8. (a) The panoramic view of the North York Moors from Crosscliffe. (b) The view from Crosscliffe with the addition of a foreground forest edge.

A view can be defined as the perception of landscape which has particular quality from a specific point.

Open views may be lost during afforestation, especially when trees are planted close to settlements, houses, roads, footpaths and the like. The landscape quality of a good view can seldom be replaced by even the most imaginative stand or edge treatment: views from or through the forest should be kept open or enhanced. Views can be 'framed' by appropriate planting, giving a greater sense of distance and scale.

Conservation of views requires:

(a) identification of public viewpoints and the particular nature of the view;

(b) design and management of foreground, and possibly middle ground;

(c) in some cases, additional landscape design and management within the view.

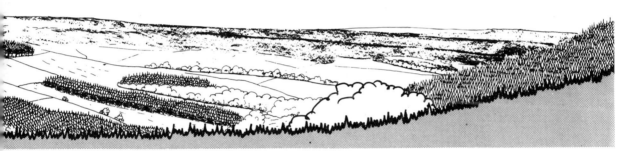

Viewpoints

Viewpoints such as settlements, recreation areas, public roads, footpaths, summits, and so on can be identified first of all from maps, and then checked on the ground. Views downhill from high ground are generally better, although views from the highest points are often disappointing. On roads and footpaths the sequence of views is important, and the direction and sideways extent of the view from key points should be identified and mapped.

Types of view

Litton (1968) identified seven compositional types of landscape, of which the first five are of particular interest here.

(1) Panorama landscape
(2) Feature landscape
(3) Enclosed landscape
(4) Focal landscape

(5) Canopied landscape
(6) Detail landscape
(7) Ephemeral landscape

These descriptions can equally be applied to views and give an indication of how views can be enhanced. The forest in the foreground and middle ground should enhance rather than compete with the quality of the view. It often 'frames' the view, but what is suitable for one type of view may not suit another. A narrow vista may frame a feature effectively, but would hide the breadth of a panoramic view. It is a safe rule that the forest framing diverse, feature, focal, or panoramic views should be kept simple so that it does not compete.

Treating types of view

Panorama landscapes are usually seen from high ground and, therefore, have little restriction from objects blocking the views. If trees are planted in the foreground, a gently curving edge underlining the view enhances the sense of limitless space more effectively than abrupt enframement.

Feature landscapes tend to be dominated by one or a few eye-catching elements. A single tree in a wide open space, or a church spire in a lowland landscape, can be as significant as a mountain. The forest should reveal and draw the eye to the dominant feature.

Enclosed landscapes are defined by identifiable open volumes, i.e. spaces which can be described in terms of 'walls' and 'floor'. Small lakes, fields, and felling coupes within the forest are examples. Such landscapes are better appreciated from above, and with an enclosing element such as a forest edge or crag behind the viewpoint, rather than open space. They can be enhanced by a frame or a light screen of tree trunks, i.e. a 'filtered' view.

Focal views usually occur in valleys and are often of high quality. They are characterized by successive ridge lines coverging and focusing the attention on lower valley slopes. A particularly high standard of design is required where these lines converge. These views are enhanced by a feature at the point of convergence.

Canopied landscapes are typical of the woodland which provides the essential overhead plane. They are best appreciated on foot; the passing motorist misses them unless they are simplified to components of clear tree boles, widely spaced, with a relatively uniform ground plane and varied canopy. They can be developed by forest management and their position is less critical than the

a)

b)

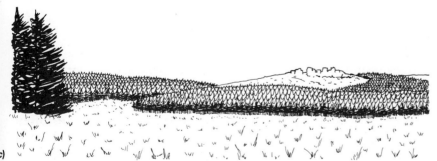

c)

Fig. 5.9. Bellever Tor, Dartmoor. (*a*) The eye is drawn sinuously through the open space to the feature even though the direct view is interrupted. (*b*) An unsympathetic interruption of the open space prevents the eye being drawn to the feature. (*c*) With careful shaping the eye can be drawn to the feature, even though the open space disappears behind the forest.

Fig. 5.10. A landscape in which the horizontal surface of the water is enclosed by the forest edge.

Fig. 5.11. A filtered view can be created where there is light, such as a water surface, field, or sky beyond. It is, however, liable to distract from the quality of focal or feature views.

previous four types. Stands which are suitable because of age, species, wind firmness, etc., should be identified as part of the landscape appraisal.

Detailed landscapes are similar in that they can be identified during landscape appraisal and developed later. They are often miniature versions of other types, with attention focused on a single feature. Their appreciation always requires a walking pace.

Fig. 5.12. A focal view in Cwm Carn, Glamorgan.

Fig. 5.13. A canopied view in the beech and oak woods of Brahan Castle, Inverness-shire, (courtesy George Dey).

Fig. 5.14. Atmospheric lighting effects at Strathyre, Perthshire.

Ephemeral landscape effects are beyond our control, tending to be related to weather conditions, lighting, water, fallen leaves, and wildlife. Ephemeral effects can be deceptive, giving an impression of high landscape quality which is quite temporary.

There are great opportunities to re-create views within, around and from the forest at time of felling. In all cases the position, direction, character, and quality of views must be assessed at an early stage of landscape survey, along with areas of forest forming part of important views which will require a high standard of design.

Water

- Water brings special qualities to the landscape and should not be obscured.

- All elements close to water features, such as forest edges, recreation sites, and structures require careful detailed design.

- Outstanding landscape quality can be achieved around water features by careful design and implementation.

Water brings movement and reflection to the landscape; still water provides a perfect horizontal plane, and a rippled surface will draw light into the landscape. The movement and sounds of waterfalls, rivers, and streams are unparalleled by other landscape

Fig. 5.15. View over a felling coupe from a forest walk at Cardrona Forest, Pebblesshire.

elements, particularly in mountains when every gully and rivulet seems to erupt with moving white water after heavy rain. Any landscape with water features is of higher quality than comparable landscapes without.

Water features, and their edges in particular, draw attention, so intrusive developments such as car parks, sewage disposal plants, etc., should be located away from the water's edge or carefully screened. A high standard of detailed forest design is required close to water; still water can be used to good effect, such as small

Fig. 5.16. A rippled surface bringing light into the landscape.

Fig. 5.17. Picturesque waterfalls are notable points of interest and valuable as diversity (courtesy George Dey).

pools seen from higher ground which reflect the sky. Surrounding dark areas of evergreens 'cutting off the lateral rays of the sun will concentrate the reflection on the deep blue of the zenith' (Crowe, 1981).

Still water will reflect the changing lights and colours of the sky, and the seasonal effects of trees. A proportion of waterside trees can be selected for autumn and winter colour.

Small streams are interesting in themselves and support a range of wildlife. Associated plant communities have their special char-

Fig. 5.18. Blue sky reflected in still waters with conifers at the edge at Tarn Hows, Cumbria.

Fig. 5.19. Water and changing light; the River Dee at Banchory, Aberdeenshire.

acter. The aesthetic quality of streamsides is high and much appreciated by visitors. The water's edge is prone to physical damage, however; protective measures may have to be incorporated in designs and some control exercised over recreational use.

Some caution is needed in the construction of new water bodies. In Britain there are legal requirements regarding the damming of streams. The aesthetic objectives should be clearly identified and it is an inviolable rule that still water should appear to lie at the lowest point in any landscape composition.

Fig. 5.20. The junction between the vertical edge of the forest and the horizontal plane of the water often appears abrupt. A gentle transition can be achieved with areas of open ground sweeping gently to the waters edge, with occasional groups of broadleaves.

Fig. 5.21. The high aesthetic qualities of streamsides is much appreciated by visitors to the forest.

Landform

● As well as their geological interest, some landform features are of high aesthetic value.

● Form, rhythm, or other aesthetic quality should be identified in each case to ensure effective conservation.

● While forest may provide an important background, interesting landform can best be seen if kept clear of trees. On steeper slopes landform detail can be reflected in the pattern of tree species.

Landform is important as an influence in the design of forest shapes and as an element of diversity. Although larger scale landform can be reflected in forest shapes, smaller scale forms tend to be lost under tree cover. These features can sometimes be reflected in the species pattern, but if the aesthetic quality of these forms is very high it may be necessary to leave some areas unplanted. Certain landform and geological features may be left unplanted for scientific reasons.

Aesthetic aspects of landform

Landform features may be of outstanding aesthetic quality. Glacial landforms such as drumlins and moraines, on the floor of many upland valleys, often contrast strongly with their surroundings.

Fig. 5.22. The horizontal lines or 'parallel roads' of Glen Roy, Inverness-shire, are relics of a former shoreline. Afforestation would hide these features and is inappropriate here.

Sometimes the form of features is of interest and repeated rhythms have a particular quality; if important, a sufficient number or grouping of forms should be left clear of trees. Smooth landform is more readily appreciated if clothed in short grass or heather rather than trees or shrubs. Dark, coarse-textured forest often provides a background against which the qualities of smooth outline are enhanced by strong contrast.

Where there are extensive areas of interesting landform, some parts should be left unplanted while others are reflected in a pattern of lighter-coloured species; the latter is more effective on slopes than on flatter ground.

Rocks, crags, and scree

- Rock features add to diversity and trees should not encroach too closely.

- This requirement is more acute where rock features contribute to the spirit of place.

- Close views of rocks and their special flora provide interest for visitors.

- Tree and shrub regeneration around rocks may require control in sensitive landscapes.

Fig. 5.23. Morainic landform in Glen Lochy, Argyll, which has been both exposed by and insensitively bisected by the new road. The forest should be reshaped to express the small scale landform which was also lost beneath the canopy.

Fig. 5.24. The form of the rock can give a particular sense of place.

Rocks provide contrasts of colour, texture, and form which are often unique to a particular scene, and which should be emphasized by design of planting and restocking rather than obscured. Crags are often lighter in colour than surrounding trees and their evident hardness adds to contrast. Rock colour is part of the *genius loci* of some landscapes; in others, it is angular form or texture which is most evocative.

Characteristic forms repeated in an otherwise smooth landscape, such as the tors of Dartmoor or the sandstone outcrops of Northumberland, are particularly important to the sense of these places. These features should be emphasized, not screened, by the forest.

Contrasts provided by rocks may take other forms, such as boulder fields or screes. Though sometimes less obvious, the landscape quality of these features can still be dramatic. In some views the rock feature is so dominant that it determines the overall approach to the forest design.

Crags which have some pockets of soil are most likely to be planted or to carry naturally regenerated trees. Such sites are rarely productive, and leaving space between the forest edge and the crag will respect the feature and increase landscape diversity.

Old quarry faces have similar visual qualities, but should be revealed with caution lest the colours of recently worked faces appears too bright. Flat, geometric quarry faces and tips may look unnatural or of an intrusive scale. One must judge whether each is an asset to reveal or a problem to be screened.

Fig. 5.25. Where outcrops are repeated at widely-spaced intervals they can be of great importance to the overall landscape. Dartmoor National Park, Devon.

Rocks can have added attraction by virtue of their texture and specialized flora, notably mosses and lichens.

Plants

- Large trees and interesting plants add to diversity, particularly in young forests, and should not be obscured unnecessarily.

- Rare species and communities often have aesthetic as well as scientific value.

- Even commonplace plants can add interest to forest landscapes, especially with seasonal changes.

- Conservation of vegetation has to take account of other landscape issues and requires careful management.

Whether occurring as natural remnants or as a result of management, trees and other plants have a high aesthetic value and may be part of the *genius loci*. Ancient trees and woods can bring a sculptural, historic, or symbolic element to the landscape. Certain plant communities have particular contrasts or combinations of leaf texture and colour, and even commonplace plants may have valued seasonal effects of blossom, leaf colour, or fruit. Ancient woodlands, riparian, and wetland communities can be particularly rich and of scientific interest.

Their effect in the broader landscape depend on visibility and scale. Large trees and extensive areas of smaller plants have their greatest impact on the longer view. Smaller details matter more near paths and recreation areas. Attractive individuals, species and communities should be conserved on areas which favour them in terms of soil, light, drainage, and fertility, ensuring that they are not obscured or eliminated by excessive tree planting or regeneration.

Old trees have a wide range of qualities which contrast with the younger forest and should not be removed unnecessarily or obscured. They provide contrasts of much coarser texture and more rounded forms of crowns, and also give a sense of continuity and heritage.

As trees decline and die they often develop great picturesque or sculptural qualities. Dead trees can create interesting textural

Fig. 5.26. Old Scots pine of the natural forest, Glen Garry, Inverness-shire, adding interest and punctuation to an already beautiful landscape.

Fig. 5.27. Old oaks in Sherwood Forest. Whether associations with Robin Hood are real or not, visitors love to see these ancient trees.

Fig. 5.28. A blanket of ivy at Ironbridge Gorge, Shropshire, creating a most unusual visual quality.

effects, but they can also appear patchy and careful assessment of scale is needed. They may easily look intrusive.

Woodland herbs such as bluebells and primroses, are a seasonal asset, well worth maintaining wherever they are visible close to paths, roads, and so on. It is best to retain the existing tree cover or to plant trees which cast light shade in spring. Common plants sometimes regarded as weeds can make a useful contribution to diversity, either occurring in large masses or in prominent positions.

Fig. 5.29. The enjoyment of a mass of bluebells under beech.

Fig. 5.30. The fiery autumn colours of beech make a powerful contrast to the evergreen forest.

Fig. 5.31. Old hedgerows are a potential source of diversity, both visually and botanically, but can look out of place in a forest setting. Partial coppicing of hedgerows and replanting additional broadleaved areas could improve these geometric shapes.

Features such as those described above should be integrated into the overall landscape design, if their aesthetic benefits are to be fully used. Diversity of these elements sometimes conflicts with other design factors such as shape, requiring a decision on how to improve shape and still retain an appropriate amount of diversity.

Archaeological sites and recreational areas

These areas are an important source of diversity and interest. Their management is dealt with in Chapter 13.

Diverse forest management

- Diversity of the forest landscape should be increased by variations in species and ages.

- Forest texture and tree size vary with age.

- Deciduous species bring variety and contrast to the forest landscape.

- All areas of contrasting age and species should be designed according to the principles of shape and scale as part of an overall landscape plan.

In contrast to the features discussed earlier, the extent and distribution of species, and ages of trees is largely under the control of forest managers. Differences in ages and species are the simplest means by which the appearance of forest stands can be diversified. Age differences show up as size differences in the short view; in middle-distance and long views age differences are expressed more as variations in texture, because of variation in height, crown shape, and crown separation. Different species add contrasts of colour and, to a lesser extent, contrasts of texture and form. Tree spacing also influences texture, but it takes wide differences to have any noticeable effect.

Fig. 5.32. Greater diversity of age and species creates a more interesting forest landscape.

Fig. 5.33. Diversity of stands as a result of varied texture of secondary forest in British Columbia.

Fig. 5.34. (*a*) and (*b*). Forest texture varies with height, spacing, and age of trees.

(*a*)

(*b*)

Fig. 5.35. The additional height of older trees introduces shadowed edges between neighbouring stands.

Variations in forest texture

Forest texture seems coarse with widely spaced trees and fine when they are close. In some natural forest landscapes variations in texture are the main source of diversity. These differences are the result of the natural distribution of trees varying according to soil conditions. Shapes, scale, and landform become more obvious where forest texture varies.

With less varied site conditions and more uniform growth, changes in texture are obtained by varying the age of different stands. Very young stands are coarse textured until the trees close canopy, remaining fine with some coarsening after each thinning. The texture of felled areas is finer until the regenerating trees reappear.

Broadleaves and Scots pine assume more rounded forms as they reach maturity, making significant contrasts with other conifers and younger trees, especially on skylines and in shorter views. Tree height also varies with age, having most impact on small scale landscapes. Differences in height introduce shadows to the woodland edge in the longer view.

Stand diversity due to spacing and management

Wider spacing may increase diversity of texture in the long view under certain systems of management such as coppice with standards. Where spacing is very even, as in some poplar plantations, the appearance can be extremely regular and formal; it may be difficult to vary spacing to any extent and, even then, such plantations may only be visually acceptable as a transition from dense forest to a predominantly open space.

Fig. 5.36. Contrast of texture between forest stands and widely spaced broadleaves in a deer park.

Fig. 5.37. Contrast of texture between broadleaved woodland and plantation poplars.

Diversity of species

Varying species is another way of introducing diversity. Strongest contrasts are between evergreen and deciduous species, the latter being lighter in colour for 7–8 months of the year, with pronounced seasonal changes, and useful contrasts of twig colours

when leafless. Levels of contrast between some broadleaved species are relatively low, as are contrasts between most species of evergreen conifers.

Fig. 5.38. Variety of evergreen conifers makes for a limited amount of diversity, but it is only significant under the best lighting conditions.

Fig. 5.39. Colour contrasts between deciduous and evergreen trees are more obvious.

Forest shapes

- The shape of woods and forest areas, margins, open spaces, and other elements influences the forest landscape more than any other factor.

- Forest shapes should be similar to those in the surrounding landscape and should follow landform, rising uphill in hollows and falling on convex ground.

- Shapes, especially species margins, can follow vegetation change at a broad scale and where there is no conflict with landform.

- Geometric shapes should be rigorously avoided, especially parallel lines, boundaries perpendicular to or following contours, and right angles.

- In any composition shapes should be combined so that they appear to overlap or interlock. Avoid belts of retained trees or different species.

- All visible lines, such as rides or compartment boundaries, should be shaped following the same principles or, preferably, eliminated altogether.

As we saw in Chapter 2, shape is the most important and evocative factor in our perception of the environment. The shapes of the following elements should, if possible, follow the principles described:

(1) all external margins;

(2) internal margins — species changes, felling areas, fire breaks;

(3) all roads and rides;

(4) all drain aligments, within the requirements of efficient drainage

Our perception of a shape is affected by its general extent and the qualities of the lines which define its boundaries. The design of the broad shape must be resolved before details of line or edge treatment are considered. All must be well resolved in terms of their *natural appearance* and their *reflection of the landscape an landform*.

In creating forest shapes it is as important to avoid mistakes as it is to follow principles. Aesthetic quality can be improved by eliminating:

(1) right angles;

(2) straight edges cutting at or near right angles to the contour;

(3) edges following contours;

(4) parallel sides;

(5) symmetrical shapes;

(6) long straight edges.

Straight lines are likely to look less unsightly where they are:

(1) short and angled diagonally to the contour;

(2) located at the lower margins and reflect valley field patterns;

(3) part of geometric layouts at the forest edge in flat landscapes where geometric field patterns predominate.

The geometric qualities of the broad shape tend to dominate the detail of lines, so the extremities of a shape should be defined first. Forest shapes should be curved gently, diagonal in emphasis, and asymmetrically balanced. Curved edges are invariably of higher aesthetic quality than straight, and where a ragged line is appropriate it looks more effective when superimposed on a generally curved shape. Shapes should be broadly related to landform with any high points positioned in the main depressions and low points near or on important spurs or ridges.

Relating shapes to landscape

The character of shapes in the surrounding landscape should influence the character of woodland and forest shapes. The angular rock formations of the Lake District may suggest angular shapes in the forest, while in smoother landscapes more flowing shapes are appropriate.

In all but the flattest landscapes, the form or shape of the ground dominates the landscape context. We have seen how the eye tends to be drawn downwards on convex ground and upwards in concavities, so if the edges of forest shapes rise more in the deeper gullies and fall more on the more pronounced spurs, a stronger relationship between the shapes of the forest and the landform is established (Chapter 2, Visual forces).

While landform is the most important cue to forest shapes, it should not be followed slavishly. Direction, symmetry, and

Fig. 6.1. Flowing shapes integrate this woodland more effectively with the smooth landform once the geometric shape of the lower margin has been amended.

balance are also important. Where there is a succession of spurs and gullies, avoid an even distribution and size of upward shapes in gullies and downward shapes on spurs. Variation in interval, size, and direction of points and asymmetry of detail is necessary. If too many sharp inflexions of edges are introduced, there is a tendency for regularity and symmetry to increase.

Landform can guide shaping at all scales. In smaller scale landscapes the detail of landform becomes more important, and edges can be shaped to follow quite subtle variations of slope.

Natural forest shapes which conflict with landform

In some mountainous parts of the world the forest shapes formed by avalanche or landslips appear to fall on hollow ground and rise

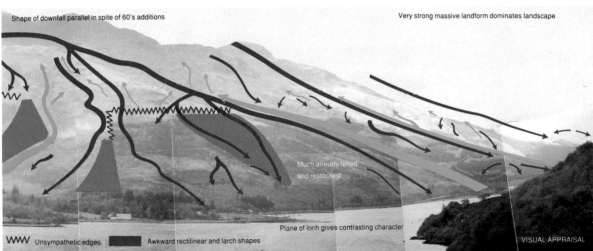

Shape of downfall parallel in spite of 60's additions   Very strong massive landform dominates landscape

Much already felled
and restocked

Plane of loch gives contrasting character

VISUAL APPRAISAL

WW Unsympathetic edges   Awkward rectilinear and larch shapes

Fig. 6.2. Landscape analysis and design proposals for part of Glen Croe, Argyll.

on spurs, apparently in conflict with visual forces as seen in Britain. Attempts to reflect the natural patterns of these landscapes, in shaping felling coupes, for example, may create visual conflict with landform.

In practice, there is conflict in a limited number of cases and other factors affect perceptions of these shapes. The eye can easily follow avalanche paths up gullies because they pass into the open mountain face above without interruption. In some places the forest margin is inflected downhill on ridges and up in hollows, especially below rock outcrops. Open space does extend down-

Fell and restock 1986
Fell and restock 1996
Fell and restock 2006
Fell 1986—do not restock
New planting 1986
New planting 1996
Long-term retention

FELLING AND RESTOCKING PATTERN

Sitka spruce   Japanese larch   Broadleaves

SPECIES PATTERN

wards into the forest in a few places where avalanche has followed ridges. In other places, the forest rising up a ridge shows a coarsening of texture due to the dispersal of trees which eliminates the potential conflict of shape and form. The upward emphasis of a forest shape sometimes ends beside rather than on a ridge, creating a shearing effect which displays tension, but not direct visual conflict.

Any of the devices above could be used where it is intended that shapes such as felling coupes should follow the characteristic pattern of these landscapes.

Fig. 6.3. Fine detail of hollows revealed by pattern of rush vegetation.

Fig. 6.4. Forest shapes falling in hollows and rising over ridges at (*a*) Tete Jaune Cache, British Columbia, and (*b*) above Lake Annecy, Hte. Savoie (courtesy Alistair Rowan).

(*a*)

(*b*)

Fig. 6.5. Despite the symmetrical appearance of avalanche paths the line of the gullies can be easily followed upwards without interruption. Banff National Park, Alberta.

The scale and position of the point where the forest shape crosses a spur have the most important effect. Conflict is worst if the forest crosses the spur in the middle third of its length. Above that the forest appears to clothe the whole feature; in the lowest third the forest seems completely dominated by landform. The forest appears less intrusive, regardless of scale, if it crosses a feature where there is a dip.

Ground vegetation shapes

Vegetation patterns can provide a model for forest shapes. They often echo landform, with shapes rising in hollows, falling in gullies, because concave ground frequently has moister and more fertile soils, and more shelter, so favouring some species while others are more competitive on drier soils of exposed ridges. The contrasts of bracken, heather, and grass can be particularly strong, and the planting of larch on bracken, spruce on grass, and pine on heather has produced attractive landscapes.

This approach should be used selectively. In translation to the scale and coarser texture of the forest, fine detail of vegetation patterns can appear fussy and the shapes should not be allowed to conflict with landform. It is safer to follow vegetation shapes for species boundaries than for external edges, as the vertical height of the trees often obscures intimate detail.

Composition of shapes: interaction of lines and edges

The layout of every boundary affects the shapes on either side. This is why forest landscapes should be designed as comprehensively as possible and all known elements, external edges, species, open spaces, felling coupes, etc., included.

Lines such as roads, rides, and racks have a cumulative effect on the forest landscape, and interact with shapes of species and external edges. Our perception of geometry is so strong that the best shaping of edges and species will be rendered ineffective if a geometric pattern of rides is superimposed.

Natural features such as streams and crags should be used, along with roads, species, and age-class boundaries, to delineate compartments. The use of rides for this purpose should be kept to an absolute minimum.

The legal boundaries of a property can include geometric shapes and straight lines, in some parts of the country going back to the original enclosure awards. Good design may well require that the forest margin is of acceptable shape within the legal boundary. It is obviously desirable that land acquired for afforestation should have legal boundaries which are sympathetic to landform wherever possible, allowing the woodland margin to be shaped closely to them and making the fullest use of the land purchased. It may also be worth acquiring adjoining land to allow improvement of poorly shaped existing margins.

Fig. 6.6. Geometric rides and larch shapes stand out strongly.

(a)

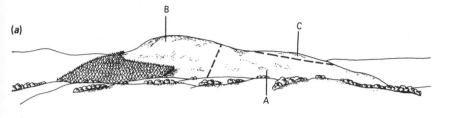

(b)

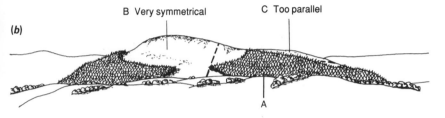

(c)

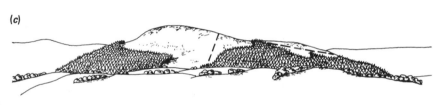

Fig. 6.7. (*a*) An additional area (A) is proposed for planting on a hill with an existing plantation. This will create three new shapes in this view, i.e. (A), (B), and (C), all of whch are affected by the design. (*b*) A moderately satisfactory design for (A) may produce quite unacceptable shapes for (B) and (C). (*c*) With (A), (B), and (C) all in mind, three satisfactory shapes can be produced.

Interlock and belts

Shapes of open space and forest, or different species shapes, can be better integrated by interlocking them so that they appear to clasp each other.

(a)

(b)

Fig. 6.8. (a) The forest and open ground are more integrated visually by the penetration of forest up the depression to the left of centre and between the knoll and open ground, left, and by the knoll pushing down between the two areas of forest. However, there are detailed shape and scale problems to resolve. (b) The integration of forest and open ground is less if the interlock is reduced. (c) An improved design with interlock at both a large and a small scale creating a strong degree of unity in the design.

(c)

(a)

(b)

Fig. 6.9. (a) Even short belts produce formal geometric effects. (b) With some evergreens brought through the edge, the composition of species layout and forest edge is united more strongly.

Continuous belts, whether planted in contrasting species for fire protection, as roadside 'amenity', or retained to screen felling, are detrimental to most landscapes. They have a geometric and formal appearance, their scale is often too small, and their effect is the opposite of interlock by dividing and disrupting landscape shapes.

Forest designs should avoid belts if at all possible. If they are essential for some functional reason, they should be discontinuous, varied in width, and with groups of varying size to give a degree of interlock.

Qualities of line and elimination of symmetry

The detailed design of edge lines has an accumulating impact on the overall appearance of the forest landscape. Most designers, including those experienced in forest design, have a strong tendency to use *regular* and *symmetrical* details in delineating forest shapes. Such regularity must be consciously avoided as far as possible if natural shapes, blending with the landscape, are to be achieved. The problem occurs most frequently in the design of upper woodland margins; symmetry can occur even with natural shapes and look artificial.

(a)

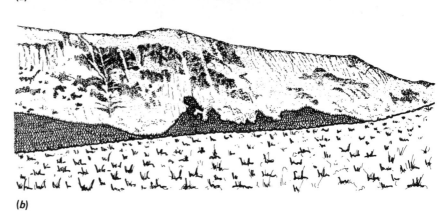

(b)

Fig. 6.10. (a) Ennerdale Forest, Cumbria. These natural shapes are the result of crop failure on unstable scree, but look awkward because of symmetry. This has occurred because the original planting had a horizontal upper margin. (b) Ennerdale Landscape Plan. Because it is unsafe to work on these unstable slopes, the design is brought to an acceptable standard by minimal felling. The most obvious symmetry is removed, but some remains in this design. (c) If more felling could be done safely, a more asymmetric design would be possible.

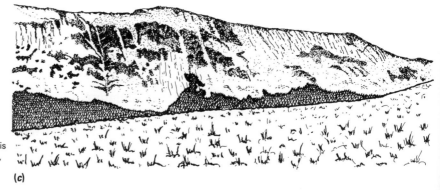

(c)

To avoid symmetry in the design of a line which rises and falls in response to landform forces, variables such as extent, interval, size, position, and shape have to be considered. Even if we design on a generally diagonal alignment, the possibilities for symmetry are numerous. The following approach may be helpful.

1. Analyse landform so that position, direction, and dominance of visual forces are known as accurately as possible. Use a contour map, and rank the visual forces in terms of size and strength of landform.

2. Establish a general shape which is asymmetrically balanced and broadly related to landform forces. Its corners may be identified as points, such as the ends of the lower boundary, where the upper margin crosses major valleys, or dips in the skyline where the upper edge might cross. Straight lines connecting these points can then be deflected in one direction or another by successive visual forces which cross them. This should be in the order of ranking established in 1.

3. When this process is complete, elements of symmetry are identified, and eliminated by:
   — exaggerating some visual forces;
   — underplaying some visual forces;
   — moving a 'peak' or 'trough' so that it is slightly offset;
   — ignoring some minor visual forces.

4. In conjunction with 3, a design of curvilinear shape is then developed.

The need for satisfactory shapes is so fundamental that it is worth making considerable efforts to follow *all* the principles in this chapter. This is not always possible, and there may be irresolvable conflicts between the principles. The following are the design priorities: all are important and the first three are critical.

1. Avoid geometric shapes, especially with vertical and horizontal edges.

2. Make curved asymmetric irregular shapes with diagonal edges.

3. Extend shapes uphill in hollows and concave ground, and downwards on ridges, spurs, and convex ground.

4. Borrow shapes from the surrounding landscape and underlying ground.

5. Interlock and overlap adjoining shapes.

External margins of the forest

External margins have the greatest visual impact of any line in the forest landscape. Contrasts of colour and texture between forest and open ground are most acute here, and design principles of shape, scale, diversity, and unity must be strictly applied to blend the forest with the landscape.

The term **margin** is used to denote the general shaping and location of the forest boundary, as distinct from the detailed design of edges. The latter is a useful addition to, but no substitute for, properly shaped margins. This chapter considers the issues and principles which affect the design of upper, side, and lower margins and skylines.

Shapes of upper margins

The upper margin of the forest is usually the most prominent and, therefore, needs careful attention. The shape should follow landform, rising in concavities and falling on convex ground. The line should reflect the quality of the landform, jagged in rugged terrain, angular in angular topography, and smooth on smooth ground.

It is often difficult to produce a diagonal direction to the line. Even when following visual forces in landform with a series of asymmetric shapes there is a tendency to create a generally horizontal line. On steep ground where an upper margin is longer than about 600 m, look for points where open ground can be brought right through the forest to the valley floor.

Although irregular shapes responding to landform are desirable, excessive irregularity can lead to problems of scale. Long protrusions can become parallel and finger-like, and appear very awkward.

Scale of upper margins

Scale is large higher up the landscape, and any element of open land lying immediately below the crest must respect the scale of the hill cap or ridge. Small slivers of open land or fringes of trees

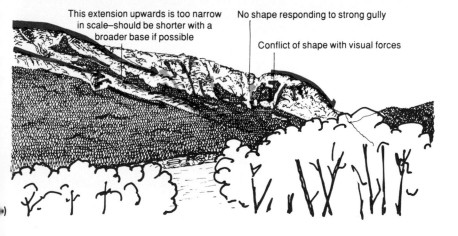

This extension upwards is too narrow in scale—should be shorter with a broader base if possible

No shape responding to strong gully

Conflict of shape with visual forces

Fig. 7.1. (*a*) An upper margin of Strathyre Forest, Perthshire. (*b*) Irregular shapes in the upper margin are badly related to landform in one area and poorly proportioned in others. Some areas are planted too high on steep slopes for safe harvesting. (*c*) Greater attention to landscape scale and visual forces produces on upper margin design with a shape reflecting the steep gully (centre) and extending less up the slope (left).

appearing over the hill are usually out of scale with the skyline. A completely planted skyline will avoid scale problems and is to be preferred. Narrow unplanted areas above the forest occur quite frequently and are most difficult to resolve in rolling topography. The choice often lies between leaving large areas of land unplanted to improve the scale, or planting more ground, on which timber yields are low.

Well shaped upper margins often involve positioning fences away from legal boundaries; this could have consequences for grazing and estate management, which should be examined in advance of design.

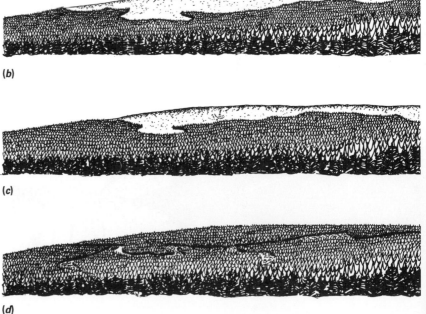

(a)

(b)

(c)

Fig. 7.2. (a) The upper edge of the forest is too straight, too horizontal, and too near the skyline. The rounded topography suggests that flowing shapes should be adopted. The geometric rides should be eliminated in favour of a more flowing pattern of open space and broadleaves. (b) A more flowing, diagonal, upper margin positioned lower down the slope improves the apparent scale of the open space, but reduces productivity markedly because a large area of potential forest would be left unplanted on these gentle slopes. (c) Reducing the unplanted area to the minimum which is visually acceptable leaves a very large area of land unplanted. (d) Extending planting to the skyline removes the narrow strip of open ground, but reduces diversity. Although the pattern of open space and broadleaves will add interest, larger areas of larch, broadleaves, or open space would be easier to design.

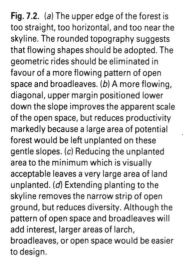

(d)

Skylines

The prominence and sensitivity of upper margins are most acute at the skyline, where the high contrast with the sky makes the qualities of landform most apparent.

Intrusive effects occur most often because scale is too small; all forest shapes near a hill top should be of adequate scale, and should reflect and not parallel the rise and fall of the skyline. In the same way that small slivers of open ground look wrong, so do small woodland elements framing sky and ridge lines.

Small scale woodland on open ground can be made to look more acceptable either by increasing its size or positioning it out of sight over the skyline. Take care lest the latter action creates similar problems on the far side of the hill.

Fig. 7.3. Small scale woodland elements close to the skyline in Sutherland and Peebles-shire.

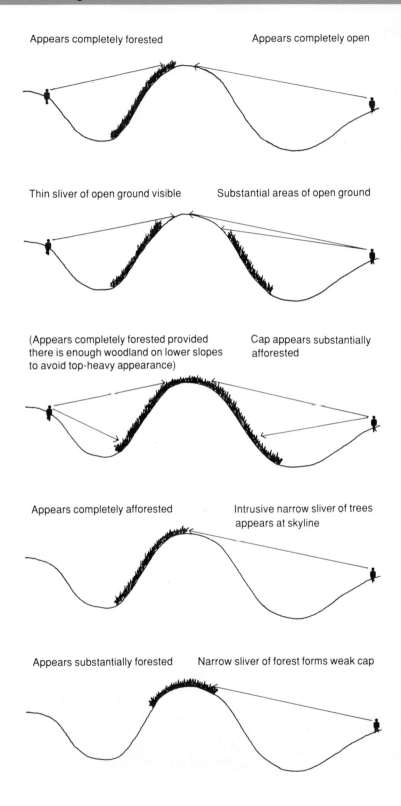

Appears completely forested

Appears completely open

Thin sliver of open ground visible

Substantial areas of open ground

(Appears completely forested provided there is enough woodland on lower slopes to avoid top-heavy appearance)

Cap appears substantially afforested

Appears completely afforested

Intrusive narrow sliver of trees appears at skyline

Appears substantially forested

Narrow sliver of forest forms weak cap

Fig. 7.4. Problems of scale at the skyline and solutions where both sides of a hill are visible. The flatter the top of the hill, the easier it is to resolve the two sides in dead ground.

(a)

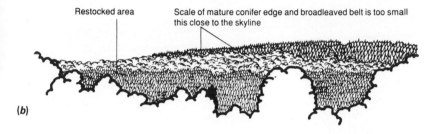

Restocked area   Scale of mature conifer edge and broadleaved belt is too small
this close to the skyline

(b)

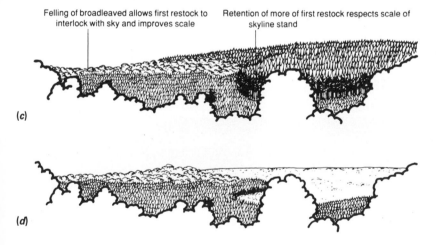

Felling of broadleaved allows first restock to   Retention of more of first restock respects scale of
interlock with sky and improves scale   skyline stand

(c)

(d)

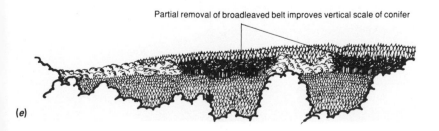

Partial removal of broadleaved belt improves vertical scale of conifer

(e)

Fig. 7.5. (a) A landscape at Tintern Forest, Gwent. (b) The felling of the middle ground area has left small scale belts of retained conifers and broadleaves at the skyline. (c) The present problems arise from earlier fellings. Ideally, they should have been phased so that a less extensive initial couple (d) would have respected the scale of the skyline in successive stages of felling. (e) Remedial action to improve the present situation is far from perfect.

Fig. 7.6. Mortimer Forest, Shropshire. A good scale of woodland on this ridge.

Fig. 7.7. Malvern, Worcestershire. The woodland on the middle ground ridge is just sufficient in scale, although somewhat weakened, centre and left. The narrow belt on the right is helped by the pattern immediately below it.

Fig. 7.8. The scale of these small woods and shelterbelts near West Linton, Midlothian, appears quite satisfactory when part of a stronger pattern of woods and fields.

Fig. 7.9. A highly intrusive fringe of forest and sliver of open space at Mortimer Forest, Shropshire. Complete planting of the skyline would improve scale.

Any forest margins crossing the skyline are particularly prominent, and the vertical edge of conifer woodlands can appear
awkward at this point. Forest margins should cross the skyline:

(1) as close as possible to a low point or saddle and away from
    summits;

(2) diagonally to both the main view and the contour;

(3) curving gently as they disappear over it.

Felling coupes close to the skyline must be of sufficient scale.
Retained fringes, groups, or scattered trees appear very intrusive
on convex skylines. At any one time, skylines should either be
substantially open or forested.

(a)

(b)

(c)

Fig. 7.10. (*a*) A forest margin bisecting the
summit of a small hill in North Wales. (*b, c*)
The illustration shows that even with
minimal adjustments of shape, moving the
margin away from the summit and towards
the saddle is an immediate improvement.

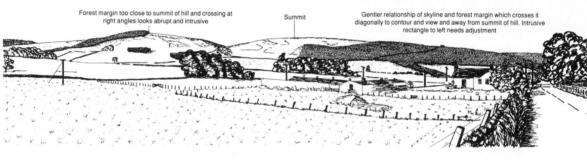

Forest margin too close to summit of hill and crossing at right angles looks abrupt and intrusive

Summit

Gentler relationship of skyline and forest margin which crosses it diagonally to contour and view and away from summit of hill. Intrusive rectangle to left needs adjustment

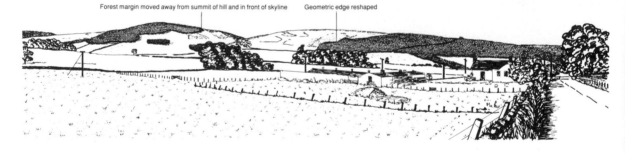

Forest margin moved away from summit of hill and in front of skyline

Geometric edge reshaped

Forest margin moved away from summit and curving gently round skyline

Fig. 7.11. Woodlands near Broughton, Peebles-shire; analysis of skyline problems and options for treatment.

Fig. 7.12. Even in the lowlands, scattered trees on the skyline have a scruffy appearance because of excessively small scale.

Fig. 7.13. The effect is less intrusive when continuous, even-aged tree canopy is maintained so that the natural shape of the skyline remains uninterrupted.

Fig. 7.14. Mortimer Forest, Shropshire. As views become longer and skylines more prominent, excessive irregularity distracts from the excellent natural shape of the skyline and gives a rather chaotic appearance overall.

Fig. 7.15. Mortimer Forest. Even quite large groups can appear intrusive in the longer view. Here a strip retention, too narrow for its skyline location, has been damaged by windthrow, leaving groups of 0.2–0.5 ha. Complete clearance is probably the best treatment.

The relative importance of landscape, financial objectives, and nature conservation may influence the planning of felling. The example in Fig. 7.16 shows several approaches.

If it is essential to reduce the scale of the felling closer to the skyline in order to favour woodland flora, an alternative solution would be long coupes running along the skyline. It is vital that the coupes vary in width and are shaped in relation to landform. Maximum coupe width can be worked out from topographic sections.

If further reduction in scale is needed at the skyline, an interlocking network of overlapping groups can be adopted. Linking these groups with areas of retained overstorey helps to strengthen the scale of the skyline. This approach is difficult to design and subsequently manage, but it does permit a smaller scale of felling

(a)

(b)

Fig. 7.16. (a) A wood on the skyline in Devon. It is proposed to phase the felling of the skyline to avoid dividing the woodland into too many horizontal layers. (b) The skyline fellings are preceded by three small coupes at the lower edge, reflecting the pattern of trees in field boundaries, introducing the first landscape changes where they have least impact, and reducing the scale of later coupes. (c) Once restocking looks well established on the first felled areas, a second felling is made on the least prominent third of the skyline, leaving a 'hedgerow' retention on the far right for balance and with the left-hand margin crossing the skyline in a gentle curve. (d) Once the second phase is well restocked the most prominent two-thirds of the wooded skyline is felled leaving a small containing group on the far left. (e) (see p. 126) If conservation of the woodland flora is important in this example, requiring shady, moist conditions, in order to survive, the felling in small groups would be better. It may cause intrusive small scale close to the skyline, however. Initially, group felling has very little visual impact, but (f) (see p. 126) at a later stage it is likely to appear fussy and to disrupt completely the natural curve of the skyline. One could opt for large coupes on the skyline and group regeneration on the lower slopes.

(c)

(d)

7.16 (*e*)

7.16 (*f*)

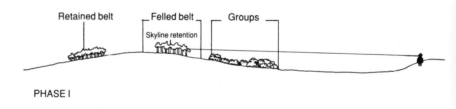

Retained belt   Felled belt   Groups

Skyline retention

PHASE I

Fig. 7.17. The wooded skyline is retained intact by felling a long screened area at an eary stage with a long retention behind it on the skyline. The first felling must be near enough to the skyline and the trees on the retained edges tall enough to screen subsequent felling there. This treatment of skyline fellings is relatively easy to plan if the wood is visible from one or two directions only.

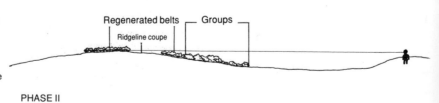

Regenerated belts   Groups

Ridgeline coupe

PHASE II

Fig. 7.18. A prominent skyline in a predominantly broadleaved landscape with conifers on the skyline. In such an area the rounded crowns of Scots pine would create stronger links with the broadleaved character of the hedgerows.

without disrupting the skyline. It is particularly important that the coupe margins should curve very gently out of sight over the skyline. This type of design is effective on areas which are seen from a number of different directions or where there are walking routes close to the skyline.

The serrated outline of conifers may appear inappropriate on the skyline of a sensitive broadleaved landscape. Greater unity with the broadleaves can be achieved if pines, particularly Scots pine, are used as they develop rounded crowns in later life.

Side margins

On open ground with no enclosure pattern, landform will tend to dominate and side margins should be shaped accordingly. A curving line is to be preferred, gently diagonal and not at right angles to the contour, with the forest below. If forest lies above the diagonal side margin, the composition can sometimes look top heavy and unbalanced.

A logical point where the forest should end is necessary, at a natural feature such as a stream, crag, or a depression. If there is no obvious natural feature, a group of contrasting species, especially broadleaves, can be planted to emphasize the end of the forest.

In enclosed landscapes where a strong hedgerow pattern extends up the slope beside the forest, straight side margins are acceptable provided that they change direction at a similar scale to the hedgerows. Where there is no hedgerow pattern, side margins should be curved, diagonal, and rising and falling in conformity with landform.

Fig. 7.19. Geometric margins in Coed y Brenin Forest are acceptable in this strongly enclosed landscape with pasture extending up the adjoining slopes.

Fig. 7.20. Side margins of the forest well integrated with broadleaved enclosure pattern. Within the forest the pattern changes so that the boundary between conifers and broadleaves is related to landform.

Fig. 7.21. Thornthwaite Forest. An unsuccessful attempt to lessen the impact of a straight side margin. Larch was planted to blend with the colour of the adjoining bracken covered slope; the latter was subsequently reseeded, introducing a contrast of colour and texture. The straight line of the forest margin with its associated differences of height and texture will continue to dominate the diffuse species boundary to the right.

HOUGHALL
COLLEGE
LIBRARY

Fig. 7.23. Broadleaves softening a conifer edge at Quantock Forest, Somerset. Their smaller size emphasizes the lower valley slopes along the streamside.

Lower margins

Although lower forest margins tend to be less prominent in distant views they often have a major impact at close range. The wall-like appearance of densely planted evergreens can appear particularly oppressive in shorter views, and the size and rounded form of deciduous broadleaves is useful in relieving this effect. Edge detail is especially important at the lower margin.

Fig. 7.22. (*a*) Darling How, Thornthwaite Forest. An intrusive side margin in a sensitive landscape. (*b*) Factors affecting the design of side margins. (*c*) Improved design, possible if additional land can be planted on the lower slope. The geometric shape will persist for some time as a light grey-brown area unless cleared of logging residues and reseeded with grass. (*d*) A diagonal line running the other way and forest above gives an unbalanced, top heavy appearance. Slash clearance and reseeding is avoided, but a large area of low-productivity plantation would be necessary to cover the prominent summit. (*e*) If there is additional hedgerow or planting to 'support' the margin, the balance may be restored. (*f*) (see p. 132) A lower line is needed to improve the side margin within the existing ownership boundaries.

(*a*)

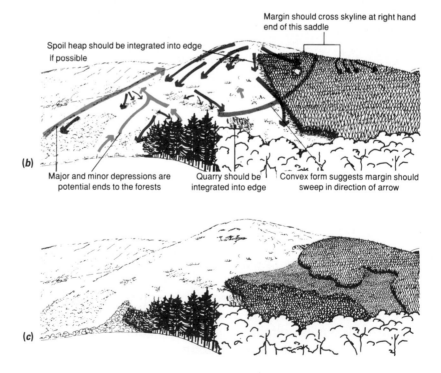

Margin should cross skyline at right hand end of this saddle

Spoil heap should be integrated into edge if possible

(*b*)

Major and minor depressions are potential ends to the forests

Quarry should be integrated into edge

Convex form suggests margin should sweep in direction of arrow

(*c*)

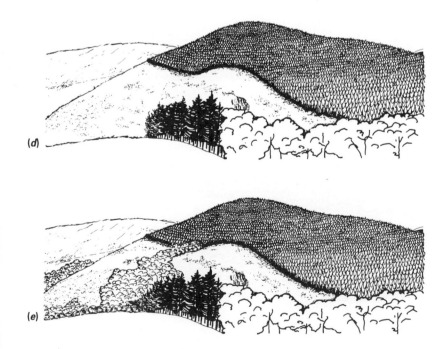

(d)

(e)

7.23 (*f*)

Fig. 7.24. (*a*) Additional planting close to stronger parts of the enclosure pattern increases interlock and reduces scale, so uniting the lower margin more closely with the agricultural landscape. (*b*) Where protrusions of the forest already exist they should not be felled so as to leave along straight line, but in such a way that open space extends into the forest.

(*a*)

Fields which if planted would reduce scale of straight lower margin and strengthen links with enclosure pattern

(*b*)

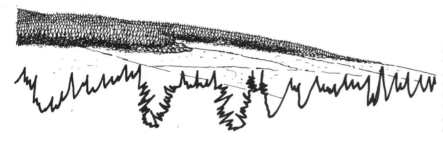

Fig. 7.25. A lower forest margin against fenced arable land, flowing down convexities and retreating in depressions. On such smooth landform the edge should be kept simple with additional species limited, at most, to sinuous broadleaved groups following the flow of the forest margin.

The lower edge should not lie too far up the slope. Unless there are some trees, in hedgerows or shelterbelts, on the lower slopes, a mass of woodland positioned too far up the slope appears to 'float' and seems unbalanced.

Where the land adjoining the lower margin consists of unenclosed ground or fenced fields, the shape of the margin should curve to follow landform. Avoid long horizontal stretches; even gently diagonal alignments make a significant improvement. If adjoining land has a pattern of treed hedgerows or shelterbelts, a more geometric shape can be adopted for the lower margin, again avoiding long horizontal lines.

It may be possible to add one or two small areas from the neighbouring fields to reduce the scale of a long intrusive lower margin and integrate the woodland more effectively with the agricultural landscape. This is further enhanced if the hedgerow broadleaved pattern is extended into the forest, especially up streamsides, gullies, and hollows, and by planting irregularly sized and spaced groups of broadleaves along the margin. Avoid continuous belts of broadleaves along the edge. Detailed edge treatment is discussed in the next chapter.

Fig. 7.26. (*a*) A lower margin following the contour looks too straight and flat. The carefully laid out upper forest margin with a diffuse irregular edge is completely dominated by the geometry of the lower edge. (*b*) Careful appraisal of convex and concave ground is an important guide to the reshaping of the lower edge. (*c*) If additional land is available, a flowing lower margin with broadleaved groups running up hollows is preferred. (*d*) If no additional land is available, an improved design can still be achieved, but weakens the scale of the forest and is still too parallel to the upper margin. (*e*) Planting the cap of the central hill solves this problem. (*f*) Open space breaking through the forest is also a solution.

(*a*)

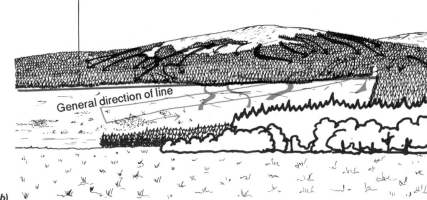

Geometric boundaries should be reshaped to give more diagonal flowing line

General direction of line

(*b*)

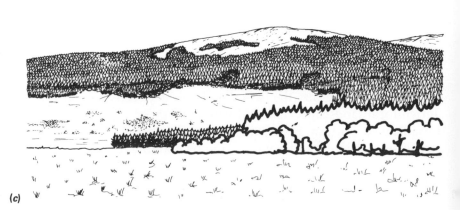

(*c*)

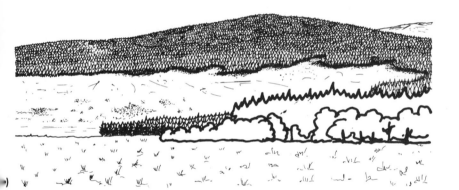

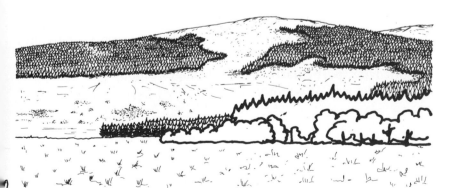

Edges

- The edges of the forest should be made to look as natural as possible and should reflect the character of the surrounding landscape.

- In wilder landscapes there should be a gradual change from the solid mass of the forest to open ground.

- Uniform edges should be varied in scale with the landscape by introducing irregular groups and changes in species, spacing and detailed shaping of the edge.

- Detailed edge treatment applied to badly shaped margins is unlikely to do much good.

There is an important visual distinction between edges and margins. Perception of the margin is dominated by its overall shape as seen in the broader landscape. In the case of edges the individual elements of trees and small groups are dominant, and detail is important. In many cases the edge detail is superimposed on a previously designed margin.

This chapter deals mainly with the treatment of external edges, but the same principles apply to internal edges and the design of the edges of felling coupes. The latter are also considered in Chapter 12. The mass of the forest contains many *ecological edges*, where there is a change of species, age, or structure; the edges considered here are, in the main, those between forest and more or less open ground. The importance of all edges to wildlife is well known, and the development of edge habitat can be readily combined with landscape design.

The opportunity should not be missed to develop visually and ecologically diverse edges at the time of planting. It is much harder to do it later, particularly where the edges cannot be thinned because of the risk of windthrow.

Design objectives

While overall shape dominates visually, the detailed texture, profile and composition of the edge affects the appearance of the forest. All such details interact and should be co-ordinated to obtain a natural effect that blends the forest with the surrounding landscape, and creates greater visual and habitat diversity.

Successful edge treatment depends on attractive forest shapes, which should be completely designed before giving attention to edge detail. There are places where even a well-designed margin can appear intrusive, especially in shorter views or when crossing skylines, and detailed edge design can do much to achieve a more gradual junction between forest and open ground. It is a waste of time, however, trying to improve poor margins by edge treatment.

The essence of edge design lie in the change from the mass of the woodland to open space and, in particular, the plane of open ground. Where elements of the two seem to overlap in a gradual change, the eye can move easily from one to the other. There is no universal treatment which can be applied in all cases, and the appearance of the edge should reflect the character of the individual landscape. A smooth edge looks right in smooth topography, while a more jagged appearance is appropriate on rugged ground. The aesthetic requirements for edges close to paths, roads, and recreation areas, where a sense of space and enclosure is needed, are different from those of the large scale upper margin of the forest.

Characteristics of natural forest edges

Certain qualities of natural forest edges can be emulated in the design of managed forests. The presence of a natural tree line indicates site conditions unfavourable for trees and a transition from forest to open moorland. The first sign of this transition is occasional gaps found below the tree line, which then develops a deeply indented edge at higher elevation with the forest edge higher in gullies, lower on ridges. Above the edge outlying trees and shrubs occur as varied groups, and scattered trees which become more widely spaced and more stunted as the forest gives way to the vegetation of open habitat.

Climate rarely limits forest growth in Britain, but various adverse conditions of poor drainage or infertility can give rise to appearance similar to that of natural forest edges. This appearance can be developed cheaply and effectively by deliberately adopting less efficient silvicultural measures on appropriate locations on the woodland fringe, as follows:

(1) planting trees directly into the uncultivated soil surface;
(2) carrying out minimal drainage and fertilizing, just enough to get the trees established;
(3) minimal weed control;
(4) no replacement of losses.

Fig. 8.1. Visual changes across a natural forest edge.

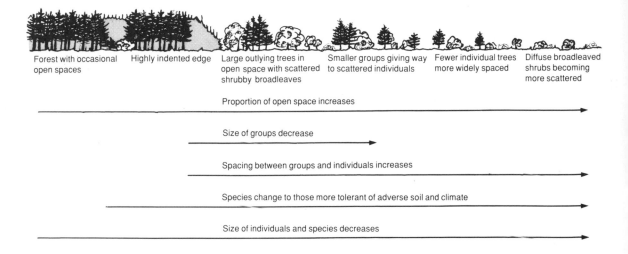

Fig. 8.2. Highly indented natural forest margin, rising uphill in hollows with outlying groups and smaller edge trees; Hudson Bay Mountain, B. C.

Fig. 8.3. Patterns of tree growth and survival similar to natural tree lines occur in areas of checked growth, especially in the middle distance, left. Wark Forest, Northumberland.

Fig. 8.4. Looking towards natural Shore pine forest (*Pinus contorta* var.) from an open bog. The decline in tree size on the edge of the bog emphasizes the natural quality of the edge. Pacific Rim National Park, B. C.

Fig. 8.5. A similar decrease in tree size as a result of poor drainage gives an apparently natural edge beside a Scots pine stand in the New Forest.

Detailed shaping of the forest edge

The fine indentations of the main forest edge affect its visual texture, and are an important link between the woodland mass and the vegetation outside. They introduce shadows to the edge, and can provide shelter and containment for recreation. These inflexions should follow the detail of landform wherever possible, and reflect the characteristic texture of the broader landscape. Their size and distribution should be irregular, so as to appear more natural and to avoid awkward symmetry. Mechanistic patterns of indentations spaced a regular distance apart must be avoided.

Tree height in the forest edge

Variations in tree height both along and across the forest edge have a significant influence on its appearance. In conifer woods the pronounced verticality of the edge profile prevents the eye moving easily between the open ground and the tree canopy. This is most marked when seen from the side. Uniform tree height along the edge creates a dark parallel band which looks unnatural and out of scale. Where tree height varies, the more diverse edge encourages the attention to move readily between open ground and forest.

A varied edge can be the result of poor growth, different ages,

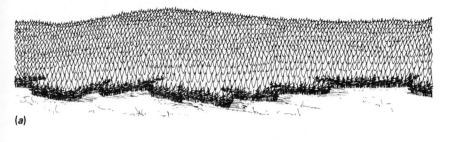

(a)

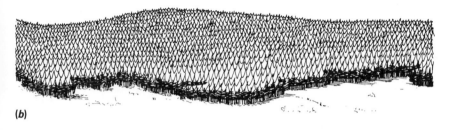

(b)

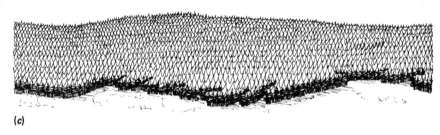

(c)

Fig. 8.6. Detailed shaping of the forest edge. (a) Indentations which are too evenly spaced and the same size produce a very regular and intrusive appearance. (b) More irregular indentations and protrusions look more natural. (c) By reducing the interval between indentations in hollows and increasing it on spurs, the landform and shape of the forest are emphasized.

or diverse species. Opportunities arise at restocking to vary tree height by age differences of larger groups. A tapered edge can be created by using smaller or slower-growing species, e.g. Scots pine planted next to spruce or fir, along with broadleaves, the latter extending outwards in a sequences of large to medium to small broadleaves and shrubs.

The face of the forest edge should not appear as a series of continuous parallel layers. This looks very artificial. The successive species should either be planted in overlapping groups or sufficiently open for different layers of trees and shrubs to mingle. This is important where contrasting deciduous and evergreen species are present.

The more gradual the variation in tree size, the more natural the edge will appear. This is more critical in wilder landscapes.

A graded edge of different sized evergreens planted occasionally is a useful means of linking the mass of the forest with open ground. Occasional gaps in broadleaves planted on the edge, where open space can flow right up to the main forest stands, is also effective. An additional benefit is that after felling the open space of the coupe can run through them and into the surrounding landscape. Smaller evergreens, e.g. Mountain pine, in the edge can help to disguise intrusive lines of roads or electricity wayleaves.

A graded edge is of particular benefit where the forest protrudes into open space and a vertical edge is seen from the side, and where edges cross skylines or ridges. Tapered edges at such points lessens the impression of awkward right angles.

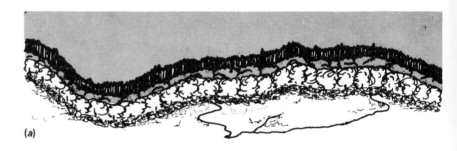

(a)

(b)

Fig. 8.7. Design of forest edge with layered profile. (a) Continuous strips of trees of different heights gives on edge a very artificial appearance. (b) Overlapping different layers in groups look more interesting and unified; appropriate for medium to small scale landscapes. (c) Different layers of similar trees more intimately mixed softens the edge without reducing the scale essential for larger scale landscape.

(c)

Fig. 8.8. Forest edge with interlocking edge and overlapping layers of various sizes of trees.

Fig. 8.9. Varied age, species, and soil conditions have given rise to different tree heights and a diverse, natural-looking edge. Inverliever Forest, Argyll.

Fig. 8.10. Planned height variation in a roadside edge at Exeter Forest, Devon. A group of pine was felled prematurely and replaced with larch.

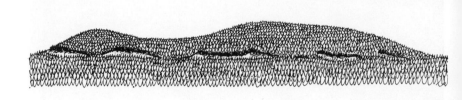

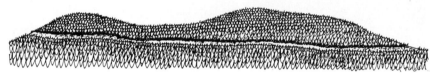

Fig. 8.11. Small conifers or evergreen shrubs can disguise the line of shadow in an intrusive internal edge.

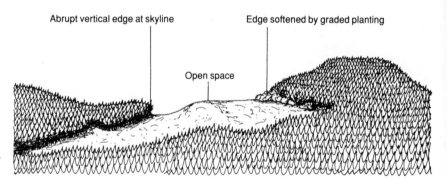

Abrupt vertical edge at skyline   Edge softened by graded planting

Open space

Fig. 8.12. Where forest edges cross the skyline awkward-looking angles with the ground can be softened by planting smaller trees and shrubs.

Groups

Where edge growth is uniform, diversity at appropriate scale can be developed by planting outlying groups. Variation in size and spacing of groups, and their proximity to the forest offers numerous possibilities for simple designs. Groups close to a forest edge provide an important intermediate scale between the broad shapes of the forest margin and outlying individual trees. Groups should be near enough to the forest to seem part of it and not appear to 'float' in open space.

Fig. 8.13. Groups of beech providing a contrast of size, form, and colour with conifers in the Forest of Dean.

Fig. 8.14. Outlying groups of birch create an interlock of open space and forest beside a public road. Strathyre Forest.

Groups should be positioned so that they appear to be a natural extension of the forest; the closer they are to the edge, the more this seems to be so. They should be placed sufficiently clear of the edge to prevent their trees developing unbalanced crowns which will appear unsightly when the edge is felled.

The general distribution of groups should reflect the broad shape of the margin, with slight indentations of the edge opposite groups.

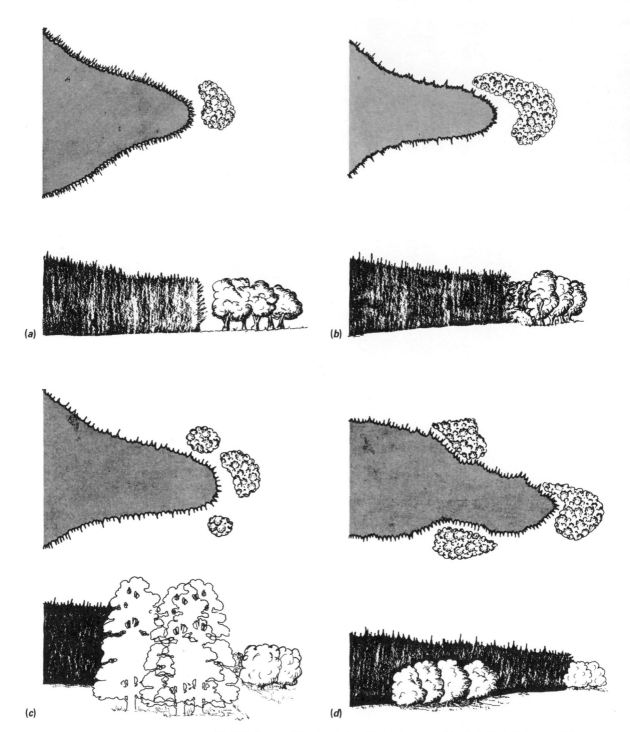

Fig. 8.15. How to unify the forest edge with outlying groups. (*a*) undesirable slot between outlying groups and the forest. (*b*) unsightly gap avoided by wrapping group around on extension to the forest. (*c*) Use overlapping groups and (*d*) slight indentations of the forest edge opposite the nearest groups.

Detailed design of groups

Once the position and extent of groups has been decided, their composition can be considered. Some landscapes will require small groups, similar to those in adjoining hedgerows and parkland. In the latter case, each group should be planted with a single species of large stature to give a strong identity, the trunks of which can in due course be pruned to reveal the ground in shadow under the trees.

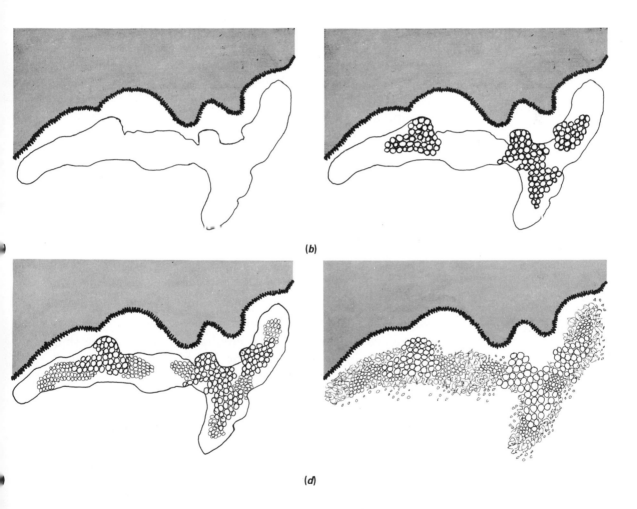

(b)

(d)

Fig. 8.16. Successive stages of design of a large group at the forest edge. (a) Plan the overall shape as part of the edge pattern, so as to emphasize the link between forest and open space; areas of largest trees are positioned asymmetrically within the group and link occasionally with the edge. (b) Areas of medium to small trees and large shrubs extending from large trees towards the convex edges of the group. (c) Remainder of the group is made up of medium to large shrubs. (d) Occasional scattered plants positioned in next smallest layer of vegetation, with small groups and individuals of the forest crop species positioned along the outside edge of the group, so that the forest appears to overlap the group. (e) (see p. 148) Elevation of a diverse tree group.

8.16 (e)

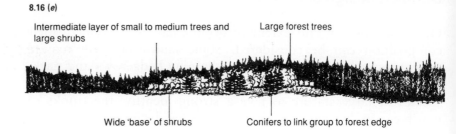

Intermediate layer of small to medium trees and large shrubs

Large forest trees

Wide 'base' of shrubs

Conifers to link group to forest edge

In other landscapes, objectives of visual and habitat diversity may require more complex designs, because the wide tree spacing necessary for development of shrub and herb layers often seems out of scale. Groups can become so indefinite that the diverse edge pattern is lost and the change from forest scale to individual tree is too abrupt. Trees should, therefore, be close enough to look like a group, but with enough open space around them to make a clear contrast. The spacing will vary with the extent of the group and size of the tree, and can be adjusted by later thinning.

A few groups situated in extensive space can appear isolated. In such cases the group looks better balanced and 'sits' more closely to the ground if it is wide at ground level and narrows asymmetrically to the top.

Belts

Belts of larch or broadleaves have often been planted at the forest edge in the past for fire protection or as 'amenity'. The latter is a misguided practice. The belt emphasizes any intrusive shape and invariably looks artificial. The same applies to retained belts beside felling coupes. Parallel shape and narrow scale of belts seem to separate the forest from the surrounding landscape rather than blending the two. Avoid creating such belts and where they already exist their appearance can be improved by breaking them up into irregular groups.

It should not be necessary to retain belts to screen felling coupes if the latter are properly shaped and in scale with the landscape. Overlapping groups are equally effective where screening of any sort is required, and do not disrupt the longer view. If possible groups should be retained on knolls with clearances in hollows, or vice versa.

(a)

Felled   Felled   Felled   Felled   Felled

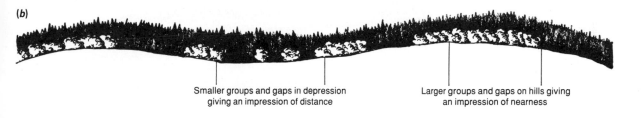

(b)

Smaller groups and gaps in depression
giving an impression of distance

Larger groups and gaps on hills giving
an impression of nearness

(c)

Fig. 8.17. Partial felling of a broadleaved edge belt: (a) on a proportional basis; (b) with larger and smaller
groups in relation to topography; (c) as in (b), but with taller species on convexities.

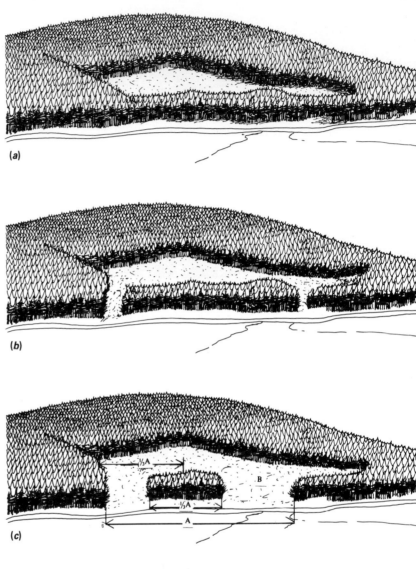

(a)

(b)

(c)

Fig. 8.18. Felling a retained belt to leave
groups screening a clear fell. (a) An intrusive
belt retained to screen a felling coupe from
the road. (b) The sides and lower corners of
the coupe are adjusted to give more
diagonal emphasis before breaking up the
belt. (c) A retention one-third of the size of
the gap and centred one-third of the way
across leaves too large a gap on one side (B).
(d) A second retention, one-third of the
length of (B) and centred one-third of the
way across, reduces the scale of the gap in
an irreguar and asymmetric design. Any
additional retentions would merely re-
emphasize the line of the original belt.

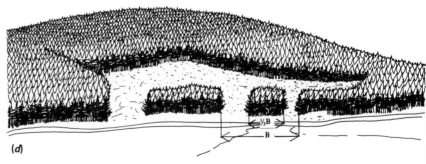

(d)

Spacing

Changes in spacing are a means of diversifying the texture of the forest edge; creating a gradual change from tree groups to open ground; establishing diverse habitat; and allowing the plane of open ground to be seen flowing amongst the trees. Areas of wider spacing should be designed in the same way as tree groups, in terms of size and frequency.

Variations in spacing have to be three to five times those normally used for any permanent visual significance; open ground quickly becomes obscured by growth of branches otherwise. Trees scattered too widely and evenly look very artificial, and wide spacing is most effectively used in conjunction with groups or areas or the same species.

Where wider spacing is to be used in sensitive small-scale landscapes, it is best implemented by an imaginative and well-briefed

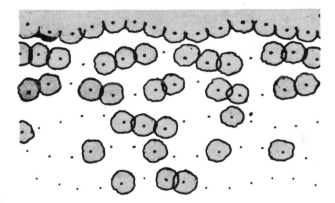

Fig. 8.21. Scattered trees should be planted so that they occasionally collect in irregular groups. Avoid a rectangular matrix of straight rows; even slight variations can alleviate the formal pattern.

Fig. 8.19. Respaced spruce at roadside in Glentrool Forest, Wigtownshire. The visibility of the ground between the trees softens the forest edge, but care is needed to vary spacing and to maintain the effect by selective pruning when branches obscure the open spaces.

planter. More systematic approaches are very labour intensive to set out. In the most sensitive areas the planting positions can be marked with bamboo canes to ensure irregular distribution.

Fig. 8.20. Poplars planted at wide spacing. Although they are broadleaves, the regular intervals create an outline which looks too formal even for this intensively farmed landscape.

Fig. 8.22. Scattered broadleaved shrubs and trees aggregate into occasional groups beside this strongly indented forest edge in Kielder Forest, Northumberland. Note the pruned trees (centre).

Thinning and pruning

Thinning and pruning increase the diversity of the edge, and allow the tree canopy and the ground surface to be distinguished. The overlap of ground plane and forest unites the two visually, an effect most obvious in smaller scale landscapes, especially close to roadsides, recreation areas, and other sensitive sites where the techniques are particularly useful. Spruce stands benefit from such treatment, but where there is a risk of windthrow, edge thinning should be carried out as early as possible to allow the development of larger root plates. Windthrow risk may be so great that it is better to adopt other measures to improve edges. Thinning combined with selective pruning is also effective in dealing with the masses of dead branches which may be revealed at the edge of felling coupes. The extent of treatment required to improve these 'brown edges' will depend on sensitivity and the degree if visual intrusion. Where conditions permit, a thinned edge to a felling coupe improves the aesthetic quality at little or no extra cost.

Pruning increases diversity by revealing tree trunks, and is most effective with species such as spruces and firs which in Britain

Fig. 8.23. An unsightly mass of dead branches, exacerbated by the regular interval between the extraction racks.

Fig. 8.24. Thinned larch, pine, and spruce at the edge of a felling coupe in Inverliever Forest.

Fig. 8.25. Occasional groups of pruned and thinned trees add to the diversity of roadside edges.

retain dead branches for many years. It is important to maintain the impression of mature trunks by felling thin whips and suppressed trees, and to thin the pruned edges to establish irregularity and avoid even spacing. Pruning should be done far enough into the stand so that unsightly dead branches are hidden in shadow of the canopy.

Pruning for appearance should be carried out to a varying height, otherwise a horizontal pruned band is created along the edge. No tree should be pruned to more than two-thirds of its total height and occasional individuals or groups with long live crowns should be left untreated to link the tree canopy with the ground. The height of pruning can be varied according to the girth of the tree or in some other irregular distribution.

Fig. 8.26. Roadside edge pruned to varying height. Kielder Forest Drive.

Chapter

9 Design of open space

Open space is a valuable element in extensive forests and as a priority for design is only slightly less important than the external margins of woodland and forest, often with comparable visual impact. Felling causes temporary open spaces in woodland, often comparatively quickly, and the design of felling coupes is a major concern of the forest manager. This subject is covered in Chapter 12.

When seen from outside in the longer view, open space in the forest appears as a shape. Seen from inside, the qualities of an open volume are more dominant. The distinction is not clear cut, as in shorter views space has both shape and volume, though one or other is usually more obvious. The best approach is to design the broad shape and woodland edge first, and then define the three-dimensional aspects of the space by minor adjustments to the edge and planting tree groups within. All changes to the design have to be checked in the appropriate long and short views.

Open space in the longer view

Scale and proportion of open space affect its design. While extensive spaces can be treated like other shapes, long, narrow or linear spaces, e.g. roads or power lines, present problems by introducing parallel lines and edges which seem to split the forest.

Numerous small spaces can appear too small in scale in long views, especially near skylines. A single gap is less intrusive provided it is enclosed by forest or near to a larger feature, such as an age or species boundary. Where several small spaces are necessary, e.g. deer glades, it may be possible to group them close enough to form a bigger entity or to be placed near larger open spaces.

Extensive spaces

We have already seen how interlocking space at the forest edge can help to blend the forest with the surrounding landscape. Extensive spaces on steeper slopes and at higher elevations, whether unplantable land, farm fields, felling coupes, or important wildlife sites,

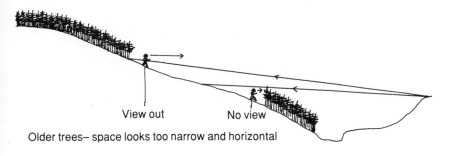

View out No view

Older trees– space looks too narrow and horizontal

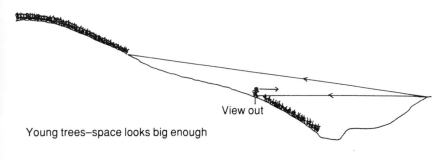

View out

Young trees–space looks big enough

Fig. 9.1. The apparent vertical height and proportions of an open space change with tree height. Better views are obtained from recreation routes positioned further up the slope.

have significant visual impact at all times of year, and the principles of forest shape in relation to landform and scale apply.

Linear spaces

Long narrow spaces beside roads and beneath power lines are often very intrusive. Position and direction of their general alignment can be as important as the overall shape, which should follow landform. The tendency of long spaces to split the forest and destroy unity requires vegetation which appears to join the forest across open ground. This can also be used to improve the shape and vary the width of the space, avoiding parallel sides. Where linear spaces pass close to the forest edge it is better to keep the intervening area largely open, rather than create narrow, small scale strips of woodland.

Power line corridors

Power lines and other service corridors require open space, the visual impact of which depends on the width of clear ground required. Narrow gaps for pipelines or low voltage power lines need

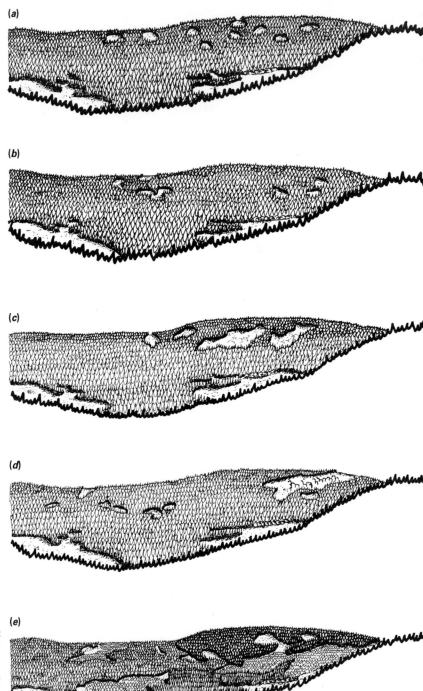

Fig. 9.2. (*a*) Small, scattered spaces look out of scale and fussy. The open space breaking through the skyline looks particularly intrusive. (*b*) Scale is improved when these spaces are placed in groups on lower slopes or (*c*) joined into larger shaped areas, or (*d*) placed close to large spaces or natural features like rocks or streams, or (*e*) tied into age or species boundaries.

(a)

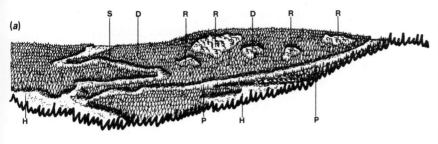

(b)

(c)

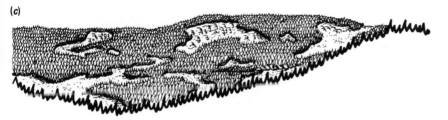

Fig. 9.3. (a) Open space is needed to reveal natural features and for management. R = rock outcrops, P = power line corridor, H = open habitat, S = unplanted streamside, D = deer management areas. (b) Linking open spaces to this degree destroys the unity of the forest, at least while the trees are young. When trees are taller some horizontal spaces may be screened. (c) Interlocking some spaces across narrow belts of forest, and some forest across narrow open space, creates a diverse, but unified landscape.

(a)

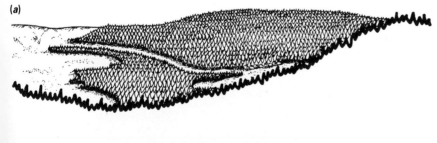

(b)

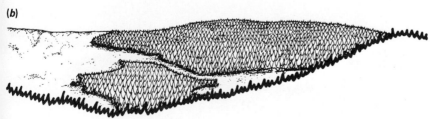

Fig. 9.4. (a) Linear space too near to forest edge. (b) Less forest improves the scale.

not be unsightly provided that they do not cut vertically across contours or follow the line of sight from important viewpoints.

Higher voltage power lines require wider corridors which are highly intrusive where they pass through forest. There are, unfortunately, many examples of misguided attempts by planners and electricity authorities to hide power lines inside forest and woodland. The resulting ugly, parallel-sided corridors are very often a much greater eyesore than the pylons carrying the line. In the vast majority of cases the lines should be routed to follow open space rather than through the forest.

Fig. 9.5. An intrusive power line corridor cutting across a ridge directly along the view from the road.

Fig. 9.6. A power line through birch scrub. Although the interlocking open space and woods disguise the corridor to some degree, the horizontal alignment (following the contour) dominates, giving a very artificial appearance.

This principle also applies to afforestation in relation to existing power lines. Open space of sufficient extent should be left for the line to pass through. If the line is close to the edge of the proposed woodland it is better to limit planting to one side and avoid the artificial effect of two parallel edges. The intrusion of power lines on the the forest landscape can be greatly reduced when the visual impact of wayleave space and towers is taken into account in the forest design.

Design of power line corridors should have regard to:

- Landscape sensitivity.
  Keep the line away from landscapes of high quality, or which affect or are seen by large numbers of people, as far as possible.

- Position.
  There will be less visual impact if the line follows depressions; in particular, the line should not pass directly over or close to a hill summit and divide it into two similar parts, but should cross the skyline where it dips to a low point.

- Direction in relation to landform and views.
  Alignment should be diagonal to the contour as far as possible and should not follow the line of sight of important views. If possible the line should be inflected up hollows and down on ridges.

Power line route planning requires the consideration of different options in terms of aesthetic merit and cost. In assessing options, sketches are used in which the design is set out, in broad terms at least. Different routes allow different standards of detailed design to be achieved, and the two cannot be separated. The example in Fig. 9.7 illustrates this interaction.

Within the forest the power line should seem to pass through a series of irregular spaces. The trees should appear to meet across the open space in some places so that the corridor does not split the forest completely. An even width of corridor is not obligatory because trees can be planted closer to the the line opposite pylons than in mid-span, where the line hangs lower and swings more. Smaller trees and shrubs can be grown closer still, as an extension of the forest edge towards the power line. This edge should be designed to create irregular spaces with irregular tree heights, avoiding severe vertical edges, particularly of conifers. The aim should be a corridor of varying character and width, swinging

Fig. 9.7. (*a*) Beattock Summit, Lanarkshire. This area is seen from the A74 trunk road which runs from just right of the viewpoint and along the base of the hill. The power line corridor is an intrusive shape and its position on the hill is very symmetrical. (*b*) Landscape planning issues which affect the position of the line. (*c*) Given the rounded landform and the power line in its present position, it is possible to design the forest to express landform, but it is difficult to avoid symmetrical shapes even if a large area is left unplanted. (*d*) More limited shaping of the plantation edges fails to avoid conflicts with landform and major problems of symmetry and scale. (*e*) If the power line is relocated over the open ground and then crosses the skyline at the lowest point, there is much less visual impact. The scale of the forest is maintained and the form of the main hill is undisturbed. (*f*) Here the power line is still prominent, but the asymmetric position on the shoulder of the hill is less awkward. (*g*) All the various options should be mapped and costs assessed before deciding on the best course of action (see p. 164).

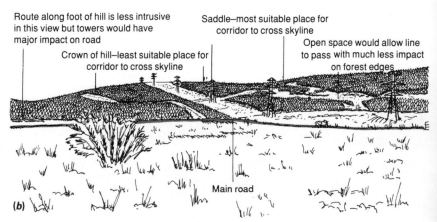

Route along foot of hill is less intrusive in this view but towers would have major impact on road

Crown of hill—least suitable place for corridor to cross skyline

Saddle—most suitable place for corridor to cross skyline

Open space would allow line to pass with much less impact on forest edges

Main road

(*b*)

(c)

(d)

(e)

(f)

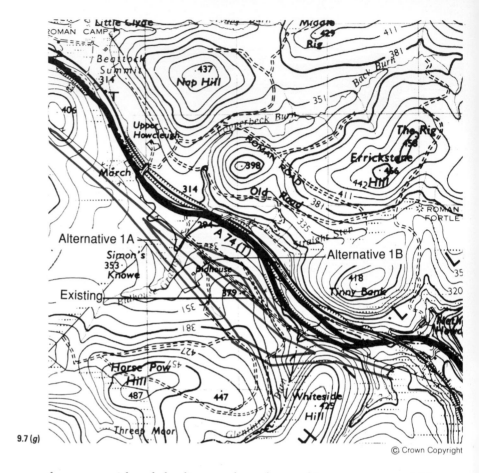

9.7 (g)

© Crown Copyright

from one side of the line to the other, taking care to avoid irregular, but parallel edges, or irregular, but symmetrical space.

Similar design considerations apply to pipelines and any other service corridors through the forest.

Linking the woodland across a corridor at time of afforestation is done by planting smaller trees and shrubby species as an extension of the productive forest edge towards the power line. Larger species should be planted next to the forest, smaller ones nearer the line. A wide range of broadleaved species can be used in woodland with a significant deciduous content, while the shrubbier varieties of *Pinus contorta* or *Pinus mugo* and indigenous evergreen shrubs, such as whin and juniper are appropriate in conifer forests. Planting the corridor with Christmas trees is not effective. They may link the forest temporarily across the space, but the effect is lost when they are cleared. Any species associated with towns or gardens should be avoided.

Where broadleaved regeneration invades the corridor space it is

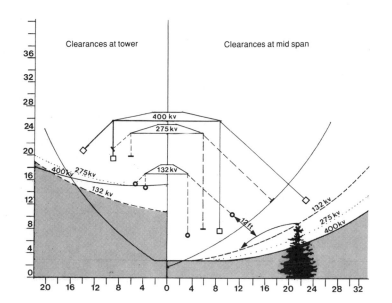

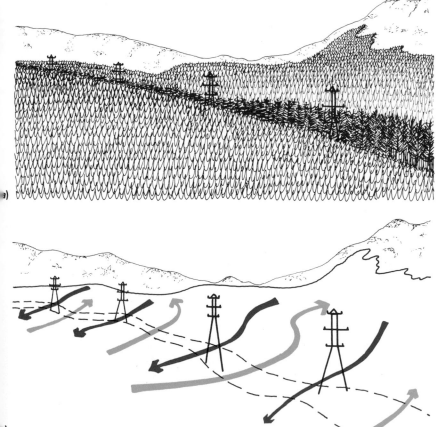

Fig. 9.8 Limits of tree growth adjacent to power lines in metres.

Fig. 9.9. Design of power line corridors. (*a*) Undesirable. (*b*) Analysis of visual forces in landform. (*c*) (see p. 166) Preferred; planting to link the forest across the space is very important when trees are small. (*d*) (see p. 166) Preferred: with taller trees the open space is more readily hidden where it is narrow.

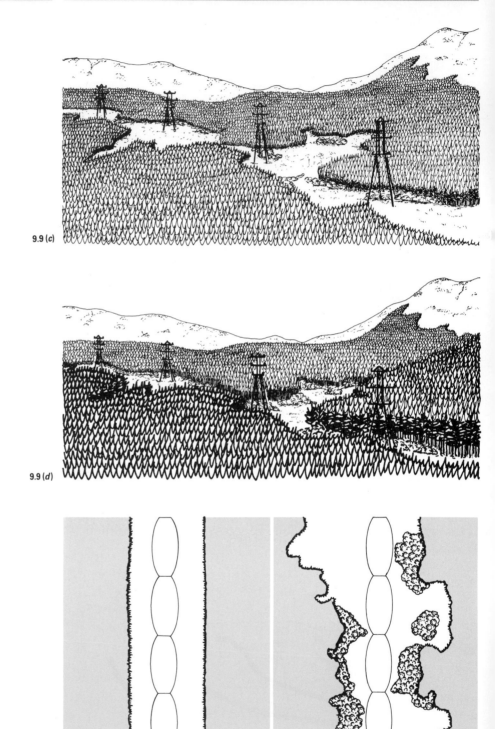

9.9 (c)

9.9 (d)

Fig. 9.10. Plan of layout; (a) undesirable (b) preferred.

(a)

(b)

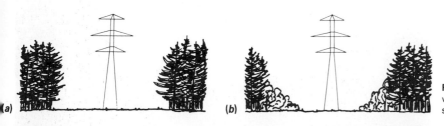

Fig. 9.11. (*a*) Section showing undesirable vertical edges. (*b*) Section through groups showing preferred edge shapes.

Fig. 9.12. A sinuous open space carrying a gas pipeline through a forest at Nunspect, Holland.

better to adopt an irregular pattern of coppiced areas than attempt any regular system of strip clearance. Areas of different age should extend diagonally from beneath the line to the edge of the corridor, and if areas of broadleaves can extend into the forest, landscape unity is improved. Completely open space should only be maintained where it makes a major contribution to landscape or habitat diversity, or where it is essential for line maintenance.

The corridor space should be widened where it meets the external margin of the forest so that it flows naturally into surrounding open ground without awkward angles.

Design of power line corridors is easier on flat ground because the forest on either side can be made to overlap more readily.

Fig. 9.13. Groups of pine and occasional birch create an irregular space beside a road and power line in Sherwood Forest.

Groups of trees within the space add occasional points of interest.

Planning of successive felling coupes should take power lines into account, as felling can temporarily increase the irregularity of corridor open space. Coupes should extend right to the edge of the corridor and be planned so that they do not appear too small or symmetrically positioned.

Design procedure for service corridors

The steps in designing service corridors in the forest are:

(1) assess alternative routes for the line, away from summits and making use of valleys, select preferred route;

(2) prepare plans and sketches to show limits of restrictions on tree and shrub planting, and to show visual forces in landform;

(3) identify areas where the productive forest edge can be placed nearest to the line so that forest appears linked across the corridors;

(4) design irregular corridor edges to create asymmetric spaces;

(5) design irregular groups of smaller trees and shrubs to link across the space;

(6) where appropriate, plan felling coupes to link with the corridor and create greater irregularity.

Design of streamsides

The management of forest streams involves the establishment of protective strips to safeguard water quality and minimize erosion. The open ground of protective strips is an important element in the diversity of riparian habitat, but the symmetry, shape, and separation of areas of forest can cause visual disunity. Landscapes with numerous watercourses can appear very fragmented.

Water management requires that at least 50 per cent of the stream should be open to sunlight with the remainder under intermittent shade from light-foliaged trees and shrubs. Periodic cutting may be necessary to maintain open ground in the face of invasive naturally-established trees of any species.

Small headwater streams must have strips of thriving vegetation at least 5 m wide on each bank. Larger streams and rivers need to have such strips two or three times as wide. To maintain strips of natural ground vegetation on this scale, forest margins of shade-casting trees must be kept back far enough to allow sunlight to reach the stream when the trees are fully grown. This is particularly important on the south-east, south, and south-west sides of streams.

Maintaining ground vegetation is vital, so heavily foliaged trees, whether conifer or broadleaved, must not be planted in the strip. Light-foliaged broadleaves such as birch, willow, rowan, ash, hazel, and aspen are suitable, according to site and locality. Some will establish themselves naturally. Alder should be used with caution and managed so that it does not cause excessive shading. There is also evidence that alder can lead to acidification of soil on certain infertile sites and is best avoided if there is concern about acidity of water.

Edges of protective strips should be shaped to echo the land-form, bringing the trees closer to the stream on convex slopes and further back on hollows. This often leaves wet ground unplanted, with benefit to stability against windthrow and to wildlife. Groups of trees should link the forest across the open space in key places which should be identified as part of the overall design. These groups should vary in size and shape, and be placed irregularly. Extensions of the forest into open space should be emphasized by planting broadleaved groups close to the edge at these points. Conversely, avoid planting groups where open space extends into the forest, as irregularity of shape tends to be lost.

Open streamside space is often useful in deer control. A varied succession of glades, between 100 and 200 m long can be created, screened from each other by tree groups which overlap the stream-side space.

Fig. 9.14. Irregular groups of broadleaves and open space create a more varied and unified landscape than continuous belts of broadleaves. Afan Argoed, West Glamorgan.

9.15 (*a*)

9.15 (*b*)

Streamside design on steep slopes

On steep slopes streams tend to be straighter, more vertical, parallel, and evenly spaced. Protective strips on either side give a very regular and geometric effect, and these continuous open spaces split the forest and disrupt unity. In many areas natural seeding of broadleaves soon fills any ungrazed open space and where this

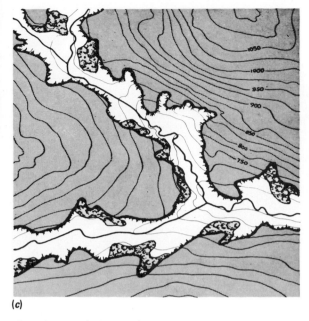

(c)

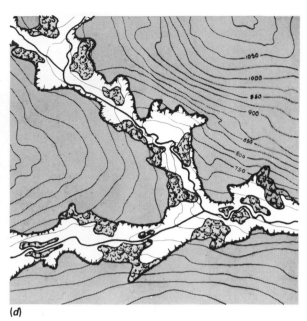

(d)

Fig. 9.15. (a) (facing page) The forest edge must be kept back from the stream to maintain bankside vegetation and so protect water quality. (b) (facing page) Planting at a uniform distance from the stream creates a symmetry which is inadequately broken by broadleaved groups in the edge. (c) The same edge extended by conifers and broadleaved groups gives a more irregular appearance, but may seem to split the forest on steeper slopes. (d) Additional broadleaved groups within the space provide greater visual and ecological diversity, and help to link the forest across the space. (e) Where streamsides are used for deer management, some groups should extend across the open space to separate adjoining glades.

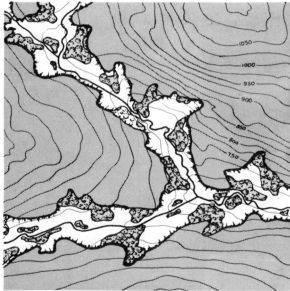

(e)

results in a pattern of evergreen blocks separated by broadleaved strips, the visual effect is unacceptable.

In such cases some small areas of conifers should be planted close to streams at selected points to create a unified landscape with a more natural appearance. When limited to a few short stretches of stream, detrimental effects on water quality are negligible.

Fig. 9.16. An irregular open space along a streamside in Clatteringshaws Forest, Kirkcudbrightshire. The distant clumps of broadleaves appear rather small in scale.

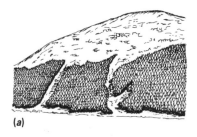

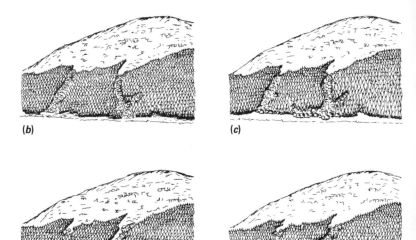

Fig. 9.17. (*a*) Continuous open space along vertical headwater streams can create awkward shapes and breaks the unity of the forest. (*b*) If the space becomes filled with self-sown broadleaves the impact is less, but shape and unity are still inadequate. (*c*) More varied width of streamsides gives more irregular shapes, but the forest is still too fragmented. (*d*) The combination of open space, broadleaved planting, and some conifers close to the streams creates a more unified and diverse pattern. (*e*) If open space at (*d*) becomes filled with self-sown broadleaves, the irregular shapes and unity are still maintained.

Lake shores and forest

Where near-vertical streams on steep slopes meet the horizontal line of a lake shore, the shapes of open space and forest can appear awkward and rectangular. Even though the basic elements of stream and lake each look natural, a gradual change from vertical to horizontal is needed when emphasized by the larger scale of the forest edge. Streamside spaces should therefore, be expanded away from the stream in asymmetric bell-mouth shapes where the forest edge meets the lake shore.

Open space around rocks

Where rocks and crags make a contribution to landscape diversity the open space around them should be planned as part of an overall pattern of open space which reflects the surrounding landscape. The amount of open space left around each will depend on its size and importance in the landscape and the steepness of the

(a)

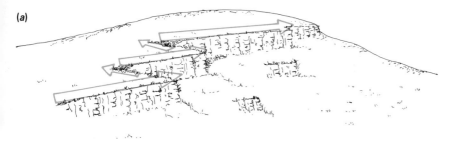

(b)

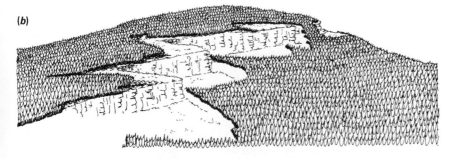

(c)

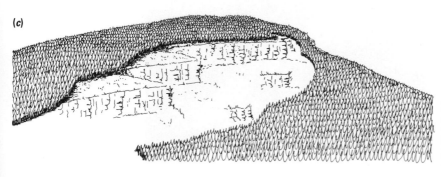

Fig. 9.18. Forest shapes reflecting patterns of rock outcrops. (a) Patterns of rock outcrops create rhythm and texture in the landscape. (b) Forest shapes reflecting the pattern of outcrops. (c) Forest shape does not reflect the pattern of rocks adequately.

slopes below. It is important to reveal individual crags in land-scapes where they are an unusual, but characteristic feature, such as on Dartmoor or in parts of Northumberland. Where crags create a strong pattern in the landscape it is more important that this should be reflected in the design of the surrounding open space than that each individual crag should be visible.

On gentle slopes and where rock features are smaller they are more easily hidden by trees and larger open areas may be needed. An idea of the influence of tree height on the visibility of such features can be obtained by drawing topographic sections. Where a feature would be unacceptably screened by mature trees there may be areas which could grow trees on a short rotation, felling them before they obscure the rock face.

The open space around rocks is designed as for other forest shapes. Where a feature is small it may be necessary to leave additional space to improve the scale.

Fig. 9.19. The open ground below this crag allows it to be seen clearly in views from the right of the picture.

Design of internal open spaces

The aesthetic principles and approach to the design of internal spaces are broadly similar, whether they contain roads, paths, water, wildlife habitat, or recreation facilities. The internal appearance of spaces may conflict with the view from the wider landscape, and the scale is quite different. Most conflicts can be avoided by first establishing the broad shape in any long views. Subsequent adjustments are usually necessary and should be checked between different views. If conflicts are irreconcilable, the longer view should be given priority because of its wider impact.

Inside the forest, open areas are seen as volumes, with sides enclosed by forest edges and the floor by the ground plane or sometimes water. The edge is most readily adjusted and has the greatest effect on our perception of the space. Conifers have a specially strong impact because of their vertical emphasis and darker colour. When grown at normal densities the sense of enclosure is very strong, and in narrow spaces appears oppressive. This effect is less marked where one can see beneath the canopy.

Aesthetic principles for internal open space

The design of the forest edge should:

(1) define a space which has a sense of enclosure without being claustrophobic;
(2) draw the eye easily from one part of the space to another while emphasizing any focal point present;
(3) create variations in the width of the space;
(4) rise or retreat on hollows and advance downhill on convex slopes and ridges.

Internal spaces can be embellished with a variety of tree and shrub groups, with the following objectives:

(1) to provide a point of interest where needed;
(2) to soften the abrupt junctions of vertical forest edges with the ground plane where they are most prominent;
(3) to balance the view as seen from important points;
(4) to add landscape and ecological diversity.

Shape of internal spaces

Compact rectangular spaces are quickly encompassed by the eye and appear uninteresting. In more irregular spaces, with different parts partially screened from each other, a sense of mystery can be created and curiosity aroused. Edges should be shaped to draw attention from side to side and through the space in a dynamic way. Any natural feature can be emphasized to provide a focus of interest.

A range of features within a space increases diversity, though if there are too many and their relative importance is unclear, the result may be confusing. Water tends to dominate a space and lines of landform focus the attention on watercourses. Forest edge design can give varying emphasis to features such as crags, interesting landforms, and large trees.

Inspiration for this aspect of design can be drawn from the English parks designed in the eighteenth century and after, in which the big house was usually the main focal point. Enclosing woodland and tree groups were used to screen some features while framing others. This created a rich landscape with a variety of features which complemented each other without competing for attention.

Scale of internal spaces

The proportions of open space are important in achieving the right degree of enclosure. Dense conifers appear oppressive if the width of the space is less than five times the total height of the trees and the effect is exacerbated in deep valleys. If the width of the space extends to much more than 10 times the height the impression of enclosure is lost.

Fig. 9.20. Designed open space at Hanbury Hall, Worcestershire a flowing ground plane and distant enclosure, with a group of *Sequoiadendron* providing a dominant feature.

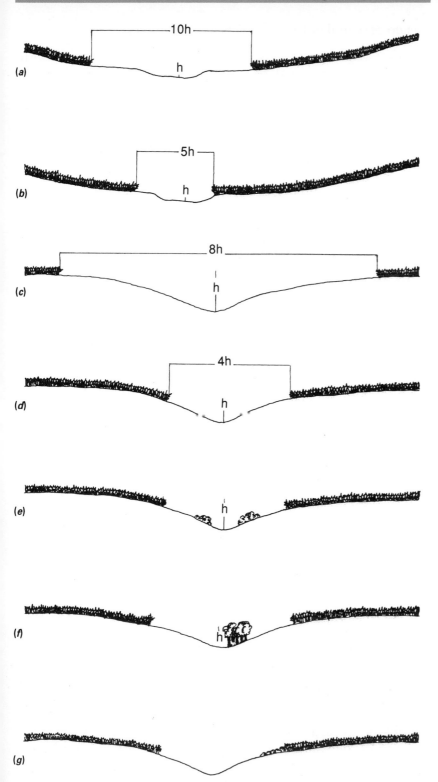

Fig. 9.21. (*a*) If width of an internal space is increased further than this, the sense of enclosure is lost. (*b*) Reducing the width of the space makes it feel claustrophobic unless the edges are modified. (*c*) On convex slopes the vertical height of forest edges may appear too small if placed further apart. (*d*) Making the space too narrow can make it oppressive. (*e*) The sense of oppressive enclosure can be relieved by reducing the apparent height of the space, or (*f*) by enclosing the viewpoint completely, or (*g*) by gradual thinning of the edge or graded planting of lighter coloured species.

Design of woodland around lakes and reservoirs

Whether water is visible to the public or not, its contribution to landscape, wildlife, and recreation should be conserved and, if possible, enhanced.

Even when disturbed by wind and waves, lakes reflect the light colours of the sky, making the latter an important element of the landscape composition. Continuous woodland around a lake tends to divide the lighter elements of sky and water from each other with a dark band. Open ground, on the other hand, can unite the composition with its intermediate colours and texture. An interlocking pattern of forest and open ground enables the eye to move easily from one element to another and comprehend them as part of a complete scene.

The appearance of a natural line where water finds its own level in the surrounding landform tends to be lost when trees are planted close to the shoreline. This line is most obvious where a promontory or peninsula extends into a lake; open space should be maintained to give views across a promontory to water beyond. Woodland should, however, reach the water's edge at some points, otherwise it will appear to float above the water on a ribbon of open ground. Groups of waterside trees which overlap with the forest edge also help to link water and forest.

Enhancing lakeside landscape

Landscape compositions which embody the relationships of forest, lake, and open space should be further enhanced by patterns of tree groups designed to make use of the reflective properties of water. Different aspects and qualities of the landscape are given emphasis at different times of day as the direction of light changes reflections and shadows in the water.

Looking towards sunrise and sunset, sun angles are low and skies can be dramatic. In these circumstances the silhouette of similar tree forms is the best complement to the sky, rather than a complicated pattern of different species. A few groups of one or two species can, therefore, be placed to enhance views to the east and west. Trees should be selected primarily for the form of their crowns and branch structure. These groups are lit from the south side in the middle of the day, so some trees with seasonal colour should be planted on the sunny side of groups nearest the water.

Looking towards the midday sun emphasizes the bright reflections from the water in contrast to dark patterns of trunks and branches. Trees beyond the water also appear dark and smooth

Fig. 9.22. Tarn Hows, Cumbria. Interlocking open space and woodland in a unified composition with water and sky. The edges of water and forest draw the eye to a focal point where an open space links water and sky (courtesy Oliver Lucas).

Fig. 9.23. Kielder Water, Northumberland. The unplanted peninsula allows the farther arm of the lake to be seen, and creates a strong interlock of land and water.

Fig. 9.24. Trees in silhouette at Wood Moss Tarn, Grizedale Forest Park.

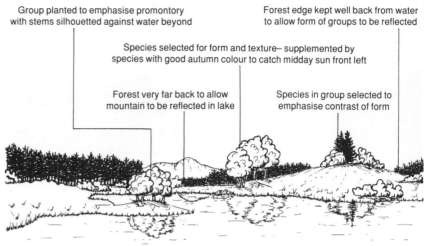

Group planted to emphasise promontory with stems silhouetted against water beyond

Forest edge kept well back from water to allow form of groups to be reflected

Species selected for form and texture— supplemented by species with good autumn colour to catch midday sun front left

Forest very far back to allow mountain to be reflected in lake

Species in group selected to emphasise contrast of form

Fig. 9.25. Tree groups with strong forms show up best silhouetted against spectacular skies. Seasonal colour on the sunny side of some groups adds interest at other times of the day.

Fig. 9.26. The dark pattern of trunks is in sharp contrast to the light reflected from the water. Landform and trees appear in silhouette in the distance (courtesy Oliver Lucas).

ground patterns appear light, so form and texture are important while colours tend to be subdued. Different layers of trees with gaps in between shows textures to good advantage.

Looking in the same directions the sun shows colour and detail to best effect. Form and texture have less impact. Seasonal colour is best displayed against a dark evergreen background. The forest edge and seasonal species need to be close to the water's edge to obtain good reflections.

Waterside design

Lake margins are important as wildlife habitat, and are best managed in a broad pattern of forest edge, scrub, open herbage, and semi-aquatic vegetation, on similar lines to watercourses. Water margin plant communities often have outstanding visual qualities and it may be necessary to keep invasive scrub in check for both wildlife and landscape reasons.

Where a water area has been created or water level raised by an embankment, always obtain professional engineering advice before planting trees or shrubs on it, or allowing them to become established naturally. The safest course is to carry out landscape measures on such structures through the medium of grasses and herbs.

Visitors have a natural desire to reach the water's edge and to walk round the lake shore. Lakeside paths should be varied by screening the lake at some points and revealing views at others; this is more interesting than a continuously open view. Wetland habitats can be vulnerable to trampling, so footpaths for visitors should either be routed away from fragile areas or proper walkways constructed.

Floating islands constructed in lakes to encourage the breeding of wildfowl are highly visible, and their shape, scale, and position has to be carefully considered. To avoid unsightly appearance:

(1) position islands as close to the shore as wildlife requirements allow and not in the central third of the lake;

Fig. 9.27. Tree groups of varying size and interval offer different views around this lockhan in Glencoe, Argyll.

Fig. 9.28. A diverse habitat of marginal plants, scrub, and forest edge which appears rather confusing.

(2) position them off-centre in bays or inlets if possible;

(3) group islands close enough to give an impression of larger scale, if wildlife requirements permit;

(4) avoid square or rectangular shapes in favour of rounded or irregular ones.

Improvement of rides

Rides were often established for access and as compartment boundaries. Modern practice is to plan road systems at an early stage, and use roads and natural features to define compartments. Existing rides must be eliminated where they are unsightly. Even those out of public view may eventually be revealed by felling. The important exceptions to this are where they are historic features,

Fig. 9.29. (*a*) Vary the direction and (*b*) vary the width, but (*c*) avoid symmetry, and (*d*) vary the scale of spaces then (*e*) increase diversity in the edge.

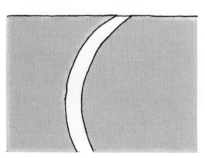

(*a*)

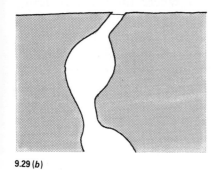 

9.29 (*b*)

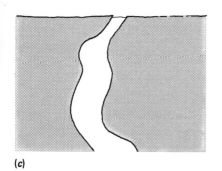 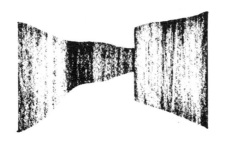

(*c*)

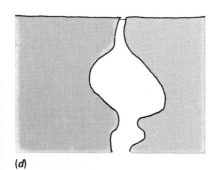 

(*d*)

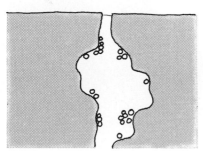 

(*e*)

in ancient semi-natural woodland, or where the value of long-established flora outweights any artificiality in appearance.

Where they are visually acceptable, rides can be an important asset for sporting and informal recreation, especially in the lowlands where they are often a traditional feature of woodland. Rides may also have considerable wildlife value; in some parts of the countryside they may be the only extensive areas of natural grassland which survive. The relative priorities of recreation and sport, wildlife, and landscape will vary from site to site.

The appearance of rides is dominated by the forest edge, and they should be of varying width, shape, and direction, designed in response to landform as are other internal spaces. They should be gently curved with a succession of spaces of varying size through which the walker passes, though the sequence should not be so varied as to disorient visitors.

The variety of wildlife habitats in rides can be increased by setting back the forest edge to allow more sunlight to reach the ground. Layout of the final edge shape is the first step. The following table gives minimum ride widths, from tree to tree, calculated for latitude 52°N. Widths are derived from the maximum shadow-free distances across rides in the growing season, i.e. between spring and autumn equinoxcs. Rides of lesser widths will not be sunny enough for a wide range of butterflies and their food plants to flourish.

Fig. 9.30. Long straight rides are daunting and unattractive to walkers.

(a)

(b)

(c)

(d)

Fig. 9.31. (a) The parallel edges of this ride define an uninteresting space. (b) A more enjoyable space flowing from one side of the ride to the other with groups of broadleaved trees and shrubs. (c) The glimpse of the distant space arouses the curiosity and draws the walker on. (d) Interesting features such as this ancient tree close to the edge of the ride should be revealed if possible; if this makes them accessible to visitors they should be rendered safe.

Height of trees (m)	Minimum ride width (m)
5	4.5
10	9.0
15	13.5
20	18.0

Rides of these widths have the following maximum possible duration of sunshine.

East–west ride
    south aspect:　5 hrs am, 5 hrs pm (10 hrs continuous)
    north aspect:　1.5 hrs before 7 am, 1.5 hrs after 5 pm

North–south ride
    east aspect:　2.75 hrs at equinox, 3.75 hrs at midsummer
               (morning only)
    ride centre:　2–2.5 hrs am, 2–2.5 hrs pm (5 hrs continuous)
    west aspect:　2.75 hrs at equinox, 3.75 hrs at midsummer
               (afternoon only)

The above figures refer to level ground. North facing slopes are less favourable and in tall woodlands in the north of Britain it is

Fig. 9.32. Progressive development of a ride for butterfly habitat and walkers. This treatment is appropriate for longer forest trails or where rights of way pass through the forest. (*a*) At year 10; (*b*) at year 15; (*c*) at year 20.

(a)

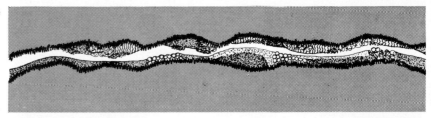

(b)

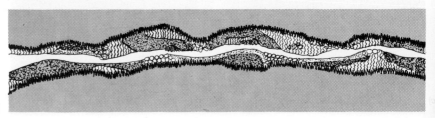

(c)

better to devote effort to improving rides running north to south on south facing slopes.

Amount of light reaching the ground can be increased by cutting 'bays' in the ride edges. These should be irregularly spaced, varied in size, and with backs not parallel to the ride. They should not be positioned opposite each other, but should be partially overlapped.

Having planned the edge shape, a path of mown grass is then designed, passing irregularly from one side of the ride to the other. Groups of broadleaved trees, tall and small shrubs, and varied areas of rough grass are then located and combined in an irregular interlocking pattern; avoid having parallel strips of trees, shrubs, and grass.

Even layers of vegetation should not be symmetrically positioned on either side of a ride, but placed as shown in the illustration Fig. 9.32 opposite.

Right-angled ride junctions should be adjusted when areas are felled and restocked. Junctions can be further improved by creating asymmetric glades.

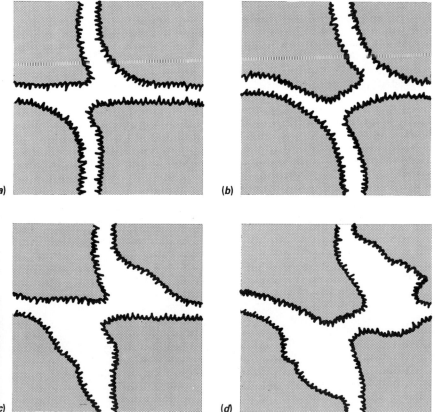

Fig. 9.33. Adjustment of ride junctions to block over-long views. (*a*) Ride junction blocking views in two directions. (*b*) Ride junction blocking views in all directions. (*c*) Ride junction with glades blocking views in two directions. (*d*) Ride junction with glades blocking views in all directions.

Much of the above can be applied to roads through the forest, though it may be more difficult to realign the space defined by the forest edge or to block long views. Where there are long stretches of road heavily used by visitors, the scale of the space may be reduced by planting occasional groups of large broadleaves close to the road, at points where the road will not suffer from surface wetness. Roadside design is considered in detail in Chapter 13.

Change of species

- One species should appear to dominate the landscape composition by about two-thirds.

- Margins between species should be designed in the same way as other forest shapes.

- Species related to ground vegetation should follow its shape at an appropriate scale and in harmony with landform.

- Mixing adjoining species at the boundary is no substitute for a well designed shape, but can enhance its appearance.

Selection of species has traditionally been a central part of the forester's art. Some fine forest landscapes have been created, particularly where the species pattern has followed ground vegetation Recent trends have been towards planting a narrow range of evergreen conifers for rapid production of timber, with larch and broadleaves planted for wildlife and landscape reasons.

The similar appearance of the main evergreen species has produced some uniform landscapes. The slight differences between these species are obvious to foresters, but not to the public, for whom the contrast of deciduous species is necessary to give an impression of diversity.

A change of species may also be introduced in order to meet the following visual objectives:

(1) to increase visual diversity;
(2) to reflect underlying variations in landform or vegetation;
(3) to introduce seasonal changes;
(4) to reduce the scale of the forest;
(5) to relieve the oppressive appearance of narrow valleys clothed in evergreen species;
(6) to solve specific design problems, e.g. planting juniper or mountain pine to link conifer forest beneath a power line;
(7) to allow more light into recreational areas.

Planning the species distribution

While introducing contrasting species will improve landscape diversity generally, the opportunity for improvement will be missed, or unattractive appearance created, unless the principles of good design are applied. There are examples in the lowlands and on more fertile upland sites where species planted in row or band mixtures are unsightly, and we have seen how belts of larch for fire protection can have undesirable shape and scale in the broader landscape.

The basic rule is that species layout should be designed within shapes of well-formed external margins and open spaces. Shapes should not be adjusted to fit the species. Poorly shaped external margins and felling coupes cannot be improved by species layout alone. Contrasting species can be selected to increase diversity, introduce seasonal change and reflect the surrounding pattern of vegetation. A broad strategy of species distribution will improve the unity of the overall design and can help to blend the woodland with the landscape.

Proportions of different species

The above objectives should aim to produce a forest landscape in which one species (usually the most productive) appears to dominate by about two-thirds. A smaller, but significant area of a contrasting species, such as larch, may be planted to reflect broad

Fig. 10.1. In predominantly larch landscapes areas of pine and spruce are valued for their contrast.

patterns of the landscape and produce useful timber. Smaller proportions of other species may be planted in irregular groups to enhance wildlife habitat, embellish edges, and provide points of interest in open space. Although the quantity of each may vary, the clear visual dominance of one species over the others strengthens the unity of the overall pattern.

It is important to remember that the two-thirds proportion concerns the landscape *as seen*. The actual proportion of productive conifers in a forest or wood may well be greater, especially in flattish or gently rolling countryside.

Visual contrasts of different species

Different aspects of a tree's appearance vary in importance when seen at different distances. Colour contrast is most important in longer views, whereas size and form of trees only appears significant in the middle distance or nearer; there is little point in trying to diversify a long view with a contrast of branch texture, if the tree colours are the same. Contrasts of form and texture are more effective when see at close range, from public roads, paths, and picnic places, for example, where the identity of different species is more obvious.

The relative colour of different species should be considered. In combination with the lighter and deciduous larch, spruces can emphasize the darker parts of a landscape; used with darker pines, spruce can act as a highlight.

Various approaches to species distribution

The broad patterns of landform or ground vegetation can be echoed in the species pattern. If vegetation has the main impact, species margins should broadly follow its changes, avoiding excessive intricacy, while a morainic landform would suggest more rounded shapes. A combination of the two could use a different species to emphasize landform and reflect irregular vegetation shapes with more indented edges of species margins. The pattern should be adjusted to avoid small isolated patches of contrasting species which may appear too small in scale and present problems of management.

There need be little or no silvicultural sacrifice in such a course of action as the landform and vegetation changes usually signify differences in site conditions which should be taken into account in any case. Even the pioneer species used for planting reclaimed

Fig. 10.2. Larch planted on rising ground and moraine with spruce in the intervening depressions echoes the landform beneath the tree canopy.

Fig. 10.3. (a) In narrow valleys extensive evergreens can appear oppressive. (b) The effect can be relieved by the lighter colour and canopy of larch.

(a)

(b)

mineral workings can be used this way, with larch and birch on knolls, and alder and pine in the hollows.

Design of species margins

Detailed design is influenced by whatever decisions were made about the broad planning of species distribution. Too much variety creates a confusing small-scale appearance and if species are mixed too intimately the differences between them may disappear in the overall pattern or texture. The visual results of planting extensive areas of intimate mixtures are rarely satisfactory, notwithstanding any ecological or financial benefits. The overall unity of the wooded landscape will be better maintained by an overlapping and inter-

Fig. 10.4. The overlapping and interlocking pattern of larch, evergreen conifers, and broadleaves at Grizedale Forest (*a*) is more unified than the horizontal layers at Tintern Forest, Gwent (*b*).

locking pattern of contrasting species. Felling coupes should be replanted in their entirety with a single species, otherwise the scale of the landscape can become too fragmented. The exception is where fellings are too large, perhaps as a result of windthrow, and a change of species would improve the scale.

When a less productive species is planted to increase diversity, there is a temptation to plant the minimum amount and to distribute it widely. This should not be allowed to conflict with the scale of the landscape. It may be better to plant one significant area than several small ones.

Shape of species margins

Whatever the reasons for introducing contrasting species, their shapes should follow the principles of good design outlined in earlier chapters, using irregular natural shapes which reflect the landform.

A well shaped boundary can be further improved by mixing the contrasting species on either side. The mixed area can be continuous or located at irregular intervals along the margin, in scale with the landscape. These practices will not disguise a geometric shape.

Fig. 10.5. The three small areas of larch in (*a*) appear out of scale. A similar total area planted in one place (*b*) looks more appropriate.

(*a*)

(*b*)

Mixing across a common margin can be done in two ways: a few groups or individuals of one species can be planted within the mass of the other; or they can both extend in opposite directions from the boundary, creating a gradual transition from one species to another. The first way can be carried out effectively by one man in the planting squad working ahead of the rest, and planting irregularly scattered groups and individuals, around which the other species is then planted. A similar approach can be adopted in areas of natural regeneration when stems are being cut out in early respacement.

Fig. 10.6. Geometric shapes still dominate the landscape even when species are mixed across the edge.

Fig. 10.7. With more irregular shapes, mixed edges can look natural and attractive.

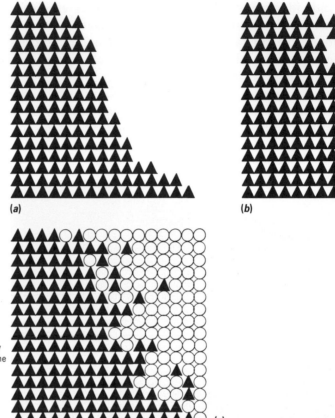

(a)

(b)

Fig. 10.8. Extending one species irregularly into another. (a) One species is planted to the marked boundary first. (b) Occasional groups and individual trees are scattered more widely from the boundary. (c) The other species is then planted.

(c)

Fig. 10.9. A small scale view where extremely intimate mixing of broadleaves and conifer gives a very pleasing effect.

It is not easy to create a truly mixed boundary which merges into pure stands in each direction, but it may be well worth the effort in a sensitive landscape. The shaped species boundary is marked out on the ground as before and a 50 per cent mixture of the two species is planted along it, alternating groups of two or three trees of one species with single trees and groups of four or five of the other. Progressing outwards in successive rows, the ratio of one species is gradually increased until planting is pure. Occasional groups of the contrasting species are added irregularly to each side to avoid the appearance of a parallel band.

In larger scale landscapes the mixture may need to extend further on each side of the boundary and the extent of the area of mixture should be assessed from the main viewpoint. The scale can be adjusted by increasing the size of the tree groups within the rows.

If the two species have markedly different growth rates, confine the mixture of scattered individuals and groups of the faster species amongst the slower. Doing the reverse results in the slow species becoming hidden and suppressed amid the mass of the vigourous species.

Mixtures

Mixtures are the least beneficial means of increasing diversity; contrasting trees are better introduced in shaped areas or groups wherever feasible, especially in sensitive landscapes.

Planting mixtures is, nevertheless, a useful way to establish broadleaves and may be the only economic method of doing so, through earlier returns from a conifer component of the stand. In landscape terms the chief difficulty lies in introducing enough irregularity to create a natural appearance without complicating management.

Row mixtures emphasize the artificial geometry of straight rows, heightened by the regular width and interval of the row. Single row mixtures of broadleaves and conifers can be easily coverted to pure broadleaves when the conifers are felled. When planted in wider bands, the strips of broadleaves persist as such after removal of conifers, often for a long time. Although mixtures of bands of three to six rows are preferred silviculturally, such mixtures should be avoided in all prominent landscapes. These objections apply equally to row mixtures of contrasting conifer species and to contrasting mixtures of broadleaves.

Shaped areas of mixtures within pure crops reduce the visual

Fig. 10.10. Planting vertical bands of alternate species looks highly unnatural. Fortunately, felling the rows of conifers will convert this area to pure broadleaves immediately.

Fig. 10.12. (a) Even when the interval and direction of rows is slightly varied and the line of species is broken in places, a row mixture looks unnatural. (b) When the small scale of the mixture is removed from the skyline, the appearance is somewhat less intrusive, but still rather unnatural. (c) When the row is curved, its appearance is less artificial, but there is still an unfortunate small scale on the skyline. (d) Curved rows kept clear of the skyline appear more natural.

(a)

(b)

Fig. 10.11. These broad bands of conifers are at least as intrusive as single rows. Felling the conifers will not remove the banded appearance.

(c)

(d)

impact because the eye is strongly attracted by the outline shape. The artificial appearance of row mixtures can also be improved slightly by interrupting the lines, and varying the species in the rows and the interval between them. The effect is still highly regular, and only when the rows are substantially curved, and an irregular outline shape defined for the mixture area, do the species begin to blend effectively. The small scale appearance of row mixtures close to the skyline can be alleviated by aligning the rows across the main direction of view or generally along the skyline but slightly diagonal to the contour.

The visual effects of **group mixtures** of one species in a matrix of another are generally easier to resolve than those of rows. Peripheral groups positioned at the edge of the designed shape help to define the outline, and the shape of individual groups is also important. Squares appear highly artificial and diamonds or hexagons with their diagonal emphasis are better. Groups distributed in lines running diagonal to the contour appear more natural than those in horizontal or vertical rows.

Variety can be achieved by varying the position and size of discrete groups, though over-large groups lose their silvicultural advantages. It is possible to distribute groups by setting out lines and spacing groups along them, though this runs the risk of producing a busy, confusing appearance. It is better to place groups irregularly within the outline shape by design on a perspective sketch, varying the interval between groups and avoiding straight line arrangements. Areas of larger and smaller groups are a useful device, using local variations in landform scale as a guide.

Further refinements are possible. Groups of mixtures can be located in a pure matrix, or pure groups in a mixture matrix. Mixtures of several species are probably best accommodated in groups of one mixture within a matrix of another. Such mixtures are complicated, not least in terms of future management, and can probably be used to best effect in small, intensively managed woodlands rather than in extensive forest areas.

Use of small groups within the forest

Within the forest small groups diversify the size of forest stands and, if carefully positioned, can provide occasional points of interest, reduce the visual impact of bad design, or emphasize or embellish attractive shapes.

The conflict of small groups with the broader scale of the upland landscape becomes more acute as their numbers increase and if

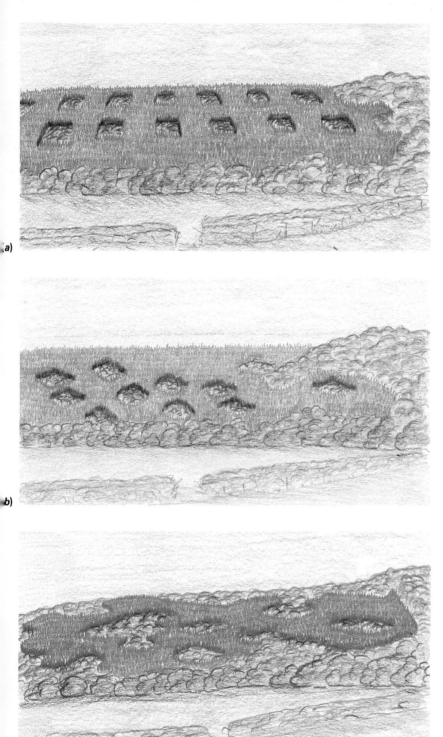

Fig. 10.13. Groups in the landscape. (*a*) Square groups in vertical and horizontal rows look very artificial. (*b*) Diagonal groups in diagonal lines appear more natural. (*c*) Aggregating groups eventually creates irregular shapes. These can be designed as described previously for pure areas of contrasting species.

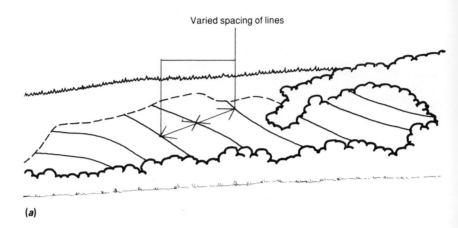

(a)

Fig. 10.14. A method of setting out irregularly positioned groups. (a) Shape the boundary of the area which is to contain the groups; then set out curving lines diagonal to the contours. The lines should be further apart at the upper edge or where they adjoin pure conifers. The direction of lines should vary progressively (roughly fan-wise) and should be a maximum distance apart of twice the length of the side of a group. (b) Starting at the outside of the shape, groups should be positioned each with one edge along a line set out as above. The distance between them should be varied, from one trees space to a distance equal to three-quarters of the length of the side of a group. Mark out the complete pattern so that the outline of the groups is visible.

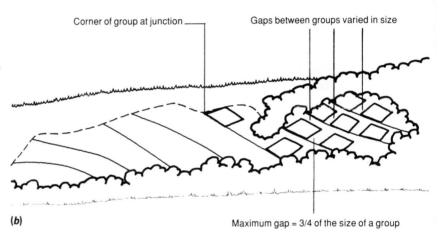

(b)

they are positioned near the skyline. They should be confined to the lower slopes and more contained valleys wherever possible. Small groups should seem part of an overall pattern which reflects the larger scale of the landscape, for example, by placing them beside the margins of larger elements such as open spaces, coupes, or contrasting species. The importance of creating this impression of larger scale is such that it may justify planting less suitable sites with broadleaves while better sites nearby are occupied by conifers.

Groups should not be planted along the margins of geometric felling coupes or other shapes as this will tend to perpetuate the intrusion. They can be positioned slightly clear of a hard edge to distract the attention until it is reshaped.

Fig. 10.15. Design of irregularly spaced diamond groups.

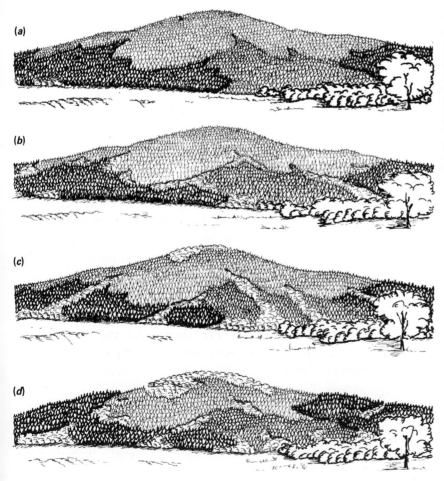

(a)

(b)

(c)

(d)

Fig. 10.16. The design of small groups. (*a*) The basic pattern of felling coupes (see Chapter 12 on how to design this). (*b*) Irregularly shaped groups of broadleaves are added along coupe edges and at junctions. Size and spacing should be irregular and they should avoid flattening out or squaring off the remainder of the coupe. (*c*) Where streams run down the face, planting the complete corridor with broadleaves tends to split coupes and disrupt design unity. (*d*) Treating the design along the water courses as outlined in Chapter 9 reasserts unity and creates a rhythm in the groups related to the coupe pattern.

Chapter

11 Visual impact of forest operations

The principles of good design should be an integral part of the planning and execution of forest work, and this chapter deals with the practical application of these principles to fencing, cultivation, and drainage, and road construction. Felling and restocking are considered in the next chapter.

Fences

The line of a fence is often emphasized by the contrast of vegetation caused by differences in grazing pressure on either side. The contrast may take years to develop, but can be very unsightly, particularly where geometric shapes result. Irregular fence lines which follow landform are a simple and permanent solution; a balance should be found between their appearance and the needs of efficient construction. Long straight stretches of fence running directly uphill or along the contour should be avoided everywhere. If this is not possible for some reason, contrasts of vegetation can be adjusted by selective management, e.g. use of herbicides, burning, or grazing. This should be applied in natural shapes extending across the fence line, creating interlocking shapes of contrasting vegetation inside and outside the fence. These treatments can be costly and technically difficut, but may be justified in the most sensitive landscapes.

In sensitive landscapes fence alignments should be designed in perspective so that:

(1) fences are positioned where they have the least visual impact, e.g. in hollows, away from skylines, and close to the woodland edge;

(2) fences run diagonal to the contours;

(3) fences follow landform, upwards in hollows and falling on convex slopes;

(4) changes in direction are at irregular intervals, in scale with the landscape, and not at right angles.

(a)

Fig. 11.1. (a) Vegetation contrast due to grazing on one side of a straight fence line can appear highly intrusive. (b) A more natural pattern of vegetation can be created by selective herbicide treatment. (c) Irregular vegetation pattern following variations in a fence line.

(b)

Areas which could be selectively sprayed to control heather and encourage grass

(c)

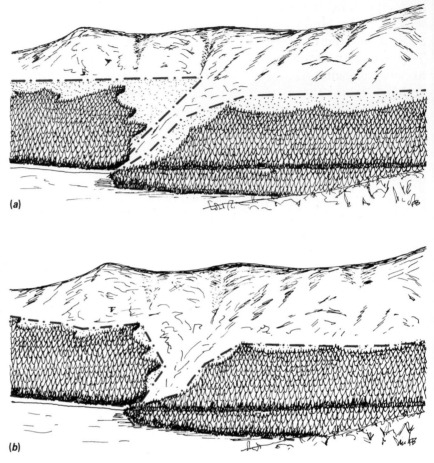

(a)

(b)

Fig. 11.2. (a) Although the forest has been redesigned to eliminate any geometric shapes, if the fence remains on its former line the grazing pattern will still show up the previous shapes. (b) Realignment of the fence as close to the new upper margin as possible eliminates the former geometry.

(5) fences either follow the woodland edge as closely as possible, or leave an intervening shape which is asymmetrical and in scale with the landscape.

Constraints on fence alignments

Sympathetic design of fences sometimes requires deviations from the ownership boundary; it may be necessary to mark the legal boundary by some other means and possibly to negotiate a token rent on the intervening area to prevent grazing or other rights becoming established. In Scotland, significant deviation from the line of a march may free a neighbour from obligations to share the cost of a common boundary fence. It is best to avoid these complications by negotiating legal boundaries in harmony with the landscape, either at time of purchase of the land or by later exchange, sale, or lease.

Fence lines should follow ground where erection is feasible, and avoid areas where deep snow gathers and allows animals to cross. The alignment should not impede grazing stock outside the forest moving to shelter in severe weather; local advice from shepherds and keepers is valuable, particularly where flocks are accustomed to move to shelter unshepherded. Sheep generally move downhill along customary routes as weather deteriorates, and fences should deflect them safely from such routes on a downward path or along the slope. Avoid fences which require sheep to move sharply uphill or to retrace their steps. They will stop, become trapped, and may perish.

Visual impact of fences

Fences seen at close range, beside roads and recreation areas, are obvious artificial structures. Deer fences are particularly intrusive at this scale. In such situations removal of fences which are no longer necessary improves the natural quality of the forest, and should be carried out as soon as possible. An alternative is to re-erect the fence inside the trees where it is out of sight.

High tensile spring steel fencing uses fewer irregularly spaced posts and blends more readily with the landscape than the traditional post and wire fence. It is also cheaper.

Various types of fence used in urban situations, such as ranch boarding, interwoven panels, industrial chainlink, and so on, are quite out of place in a woodland setting. If it is necessary to have fences at all around forest buildings, these should be of a simple post and rail type for larger enclosures, and vertical spaced boarding for smaller ones.

Cultivation and drainage

Often the first major landscape change in the course of afforestation, the process of breaking the ground for cultivation and drainage is perceived to be symbolically violent and destructive of a familiar cover of vegetation. The appearance of straight evenly spaced lines is highly artificial and quite distinct from the finer texture of agricultural ploughing. The patterns produced by mound cultivators and scarifiers may appear less regular than those of forest ploughs.

(a)

Fig. 11.3. Complete cultivation reveals natural patterns of soil changes (a) but with furrow ploughing the regular lines dominate visually (b).

(b)

Design problems

The most immediate contrast is between ploughed and unploughed ground, so areas of ploughing should follow the proposed shapes for planting as accurately as possible. Areas which are to be left uncultivated, e.g. steep slopes, should be irregularly shaped to follow landform even if they will eventually be hidden by trees. Routes left unploughed for the passage of cross-country vehicles distributing plants, fencing materials, etc., should be kept to a minimum width and number. They should follow natural breaks in slope and lower, less visible ground and should be later planted.

Ploughing appears most intrusive where furrows run straight in one direction over a large area. This is more common on gentle slopes than on steep ground where safe working requires more frequent changes in direction. To avoid geometric effects, ploughed furrows should run precisely downhill in gentle curves, with more definite changes of direction at natural changes in slope. Slight

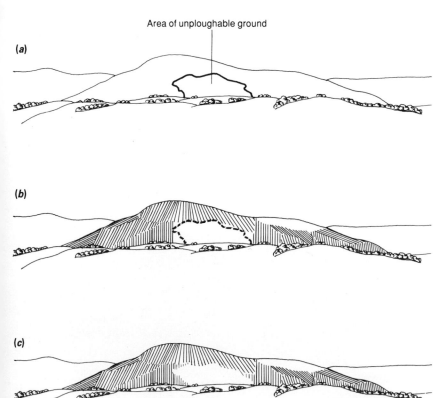

(a)

(b)

(c)

Fig. 11.4. (*a*) Areas of ground which are to be left unploughed should be identified on a sketch. (*b*) Unploughed areas should not be adjusted into square, geometric shapes, but (*c*) should follow landform with rising diagonal margins.

Fig. 11.5. Ploughing pattern revealed by tree rows against the snow cuts straight across gently rolling landform on a vast scale. The unploughed rides are particularly intrusive and will be visible for many years. Forest of Ae, Dumfries-shire.

curves can appear quite pronounced on gentle slopes because views are often foreshortened.

More changes in ploughing direction are necessary in smaller scale landscapes with close views, especially at roadsides. Furrows running at the same angle to long stretches of motor road can distract drivers. Frequent changes of ploughing direction relieve this distraction and give a more human scale.

Design procedure for ploughing in sensitive landscapes

Design can be carried out quickly on a contour map at 1:10 000 scale or thereabouts. Such maps show broad landforms more clearly than can be seen on site. This procedure will give a general guide for on-site use, at little cost. The increase in time for site supervision and setting out can be minimized by following the principles generally, rather than attempting to lay out the design over-precisely.

- Identify and record areas of steep ground where ploughing direction will be determined by position of watercourses and considerations of operational safety.

Fig. 11.6. More varied topography requires the plough operator to change direction more frequently and a smaller scale irregular pattern is produced. Strathyre Forest.

Fig. 11.7. Slightly curved furrows and occasional changes of direction reflect the forms and scale of the landscape. Whitelee Forest, Ayrshire.

- Draw on the map, in curving shapes, the main landform areas identified by significant changes in contour spacing and direction; also detailed landform visible at roadsides and in close views. Temporary access routes should follow the boundaries of these areas in low ground wherever possible.

- Record the precise direction of slope at about 50–100 m intervals along contours.

- Within the larger areas outlined, sketch a gently curving furrow alignment running down the slope which provides maximum length of furrow over the greater part of the area.

- Within the smaller areas the furrow alignment should follow the longer side of the shape, if slope permits and drainage is adequate.

Drains

In most cases the alignment and gradient of drains is determined by the need to avoid erosion while removing water efficiently. Maximum gradient is 3 per cent and this frequently results in curved shapes. On some gentle slopes straight alignments can be produced, but should be avoided wherever possible, even in the short cut-off drains used to prevent ploughed furrows carrying too much water.

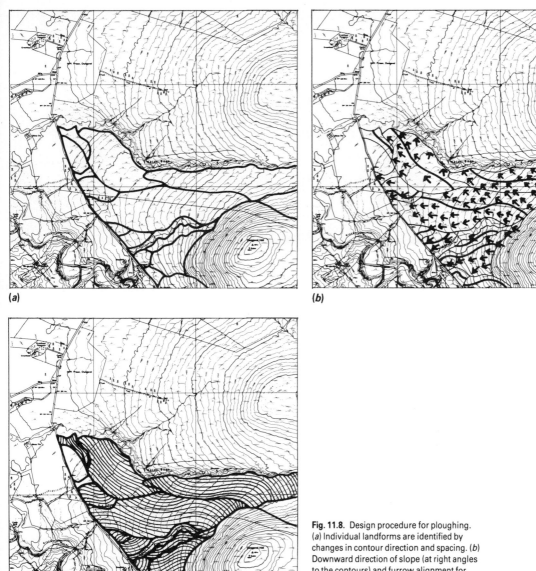

(a)

(b)

(c)

© Crown Copyright

Fig. 11.8. Design procedure for ploughing.
(a) Individual landforms are identified by
changes in contour direction and spacing. (b)
Downward direction of slope (at right angles
to the contours) and furrow alignment for
each is added. (c) Sketch plan of ploughing
pattern is then prepared.

Fig. 11.9. Straight drain lines (centre) should be avoided wherever possible and even short cut-off drains should be curved.

Other methods of cultivation

Recently scarification, mounding, dolloping, and ripping have become viable alternatives to ploughing on many sites. Generally, their impact is much less than that of ploughing. However, straight rows of mounds or dollops can be as intrusive as straight plough furrows. The inherent flexibility of all these methods should enable more irregularity of alignment and spacing of individual planting sites than is possible with ploughing.

Forest roads

Roads were considered earlier in the context of the design requirements of linear space. The intention is not to disguise the road and other structures, but to reflect the scale and forms of the landscape in their design, and locate them so as to minimize their visual impact within the technical limitations of the harvesting road network. Where roads are planned and constructed to take account of environmental, as well as operational needs from an early stage, any additional costs are usually modest. Mistakes can be very expensive to put right.

Roads crossing unplanted ground or young woodland are highly visible in the broader landscape. Although roads constructed through older trees are less obvious, they are later revealed by felling. The geometry of the road, bare areas of fill, large scale cuttings, and bridges and culverts all appear artificial, especially in close views. Where roads cross open ground, such as access roads across farmland, the visual effect should be considered prior to negotiations for the right of way.

Route selection

Forest roads should be unobtrusive. Access roads in agricultural landscapes can be hidden behind hedgerows, walls, or in hollows screened by landform. Avoid, if possible, small-scale landscapes with strong *genius loci*, and roads should always be kept well clear of archaeological sites and wildlife habitat of special importance. Focal views, water edges, and waterfalls and watercourses of particular quality should be crossed at the least visible point. High standards of design and construction must be achieved in these sensitive locations.

The visible parts of the road should be in scale with the landscape. Roads should not run close to the skyline for long stretches and should cross skylines as near as possible to the lowest point

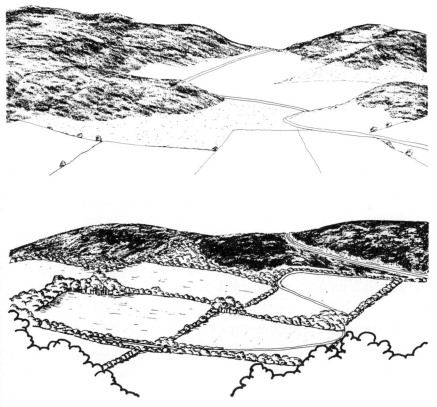

Fig. 11.10. Approach roads through agricultural land should be planned so that they are hidden by natural landform, hedgerows, or walls.

Fig. 11.11. A concrete bridge located intrusively in a landscape of fine *genius loci*.

or in a slight hollow. Steeper slopes and hill summits should be avoided, as should large areas of cut and fill in narrow valleys. The general alignment should be diagonal to the slope and not horizontal, as far as technical constraints permit.

Where there are a number of possible routes they should be sketched from main viewpoints and compared.

Landings and turning points should be sited, as far as possible, where natural gradients provide space, and not positioned on prominent spurs or ridges. This is important on steep ground where large areas of fill are needed.

Shape of road lines

Like other lines in the landscape, the shape of a road affects its appearance. The line should curve gently and blend with the landform, inflecting downwards on convex slopes, and rising slightly in hollows and valleys. The latter is important in ensuring that roadside drains, like other forest drains, do not discharge directly into natural watercourses, but empty instead into seepage zones which reduce the risk of sediment entering the stream system.

Visual impact of cuttings and spoil

Cuttings and areas of spoil produced by road construction on slopes often appear ugly because of their geometric shapes, light raw colours, and large scale. Alignment planning should obviously

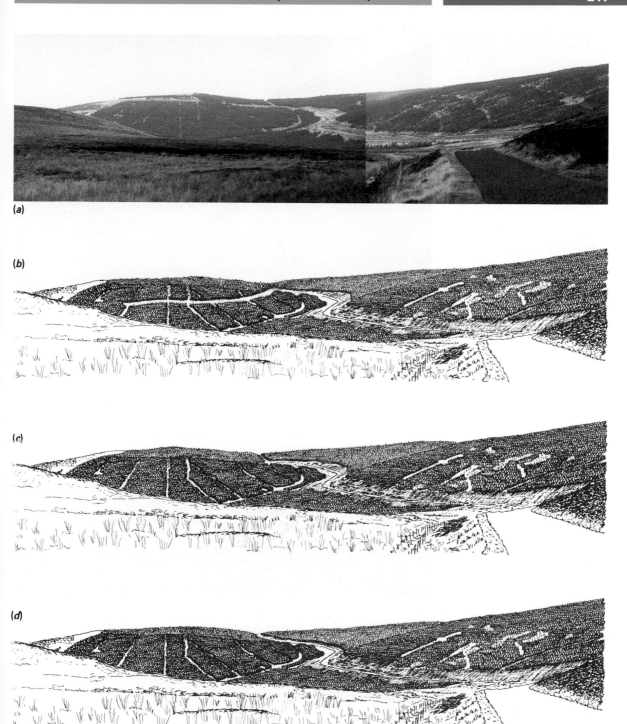

Fig. 11.12. (*a*) Although the alignment of the road follows landform well, it is out of scale with the skyline to the left. (*b*) It should either fall downhill again or (*c*) stop in the depression or (*d*) pass over the skyline at the low point.

Fig. 11.13. A forest road which curves from side to side, and up and down hill.

Fig. 11.14. A road alignment rising slightly in hollows and falling on convex ground.

minimize cut and fill to reduce costs, and it also reduces unsightly appearance.

In sensitive landscapes, road cuttings should be made less intrusive by imitating a natural profile. Where rocks are soft, cuttings should be made with rounded banks and tops sloped off to prevent the formation of overhanging turf. This, with dark shadows underneath, makes an unsightly and artificial black line parallel to the road. Cuttings in harder rock should avoid precisely even faces. Strong irregular shapes more akin to natural rock outcrops

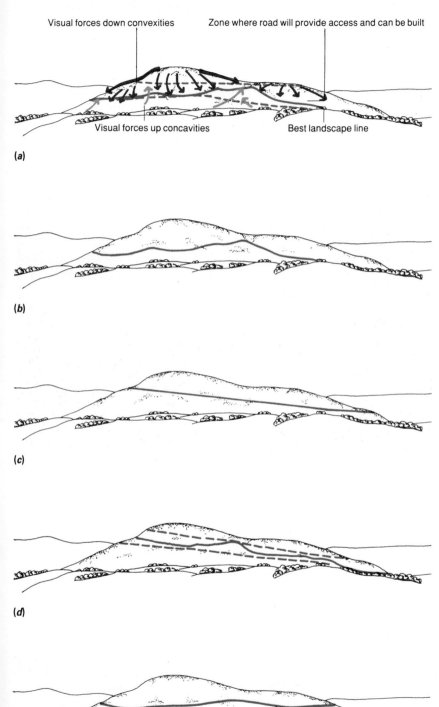

Visual forces down convexities   Zone where road will provide access and can be built

Visual forces up concavities   Best landscape line

(a)

(b)

(c)

(d)

(e)

Fig. 11.15. Road line shapes in the landscape. (*a*) Visual forces, road corridor, and road line shape. (*b*) Good diagonal line following landform. (*c*) Diagonal line, but too straight, i.e. gradient too even. (*d*) Diagonal line following landform. (*e*) Line follows landform, but is too horizontal generally.

should be the aim. The natural bedding of the rock can show through, introducing variations of light and shadow, and providing ledges where debris can collect and plants become established. Similar measures apply to borrow-pits and quarry faces.

Re-establishing vegetation on spoil can be expensive and technically difficult, and good design will produce more natural patterns and save money by identifying the most cost-effective areas for treatment. Seeding spoil with grass may simply turn an intrusive shape from grey to bright green, and unless the surrounding vege-

Fig. 11.16. (*a*) A straight road line produced by a precisely even gradient. (*b*) Analysis of visual forces along a proposed road alignment. (*c*) Road line adjusted to reflect landform.

(a)

(b)

tation is identical in colour, the effect is unlikely to be satisfactory. Irregular shapes of darker vegetation, e.g. heather, bracken, gorse, and willow, which extend across the spoil and into the natural herbage, are far more likely to be effective. An interlocking pattern of shapes will help to integrate the road line with the landscape.

Where spoil is too coarse and open for plants to grow, occasional pockets of finer material should be incorporated so that turf, bracken rhizomes, shrubs, or trees can be established. Turf and soil stripped from the site should be used as a source of vegetation; if there is no space for storage, these materials can be transported directly from where they are currently being stripped to completed areas of fill.

Tree shelters

Planting trees in rigid plastic tubes accelerates early growth and gives protection against browsing animals. The tubes are unsightly and this has to be balanced against the silvicultural advantages. They are best used for establishing small groups and individual trees; on large areas it is usually cheaper to exclude animals by temporary fencing and promote rapid early tree growth by correct herbicide treatment.

The main visual disadvantage of tree shelters is that they introduce numerous artificial elements into the landscape and reduce its

natural quality. Their reflective surface makes them highly visible, even when subdued colours are used, and their translucence makes them conspicuous against the shadows of a forest edge. Both the cylindrical and square-section forms are highly geometric.

Adverse effects can be lessened by careful choice of colour.

White	Very intrusive visually, should not be used.
Emerald, mid- or blue-greens	Appear garish against natural vegetation, avoid.
Brown	The most widely applicable colour; soft deep russets better than yellowish ochreous browns, *recommended*.
Olive greens	Relate well to grass greens, but can be garish; most effective when used as contrasting groups in areas of brown shelters.

The form of tree shelters is only seen in close views, e.g. beside roads and paths. Cylindrical shelters should be used in these locations because their form is less geometric.

Large areas of tree shelters are ugly, especially on flat or gently sloping ground where the effect is of almost continuous plastic. This is not alleviated by using a brown colour, but the large scale can be reduced by changing the colour to olive green in selected places. On steep slopes groups of shelters in brown and olive should be laid out in irregular shapes in scale with the landscape.

The artificial appearance of shelters is unneccessarily emphasized by planting in a precise geometric grid. There is no need for this.

Fig. 11.17. White tree shelters are very visible and unsightly.

Trees can be found with ease, and should be planted on the best available microsite within the general constraints of the required stocking density. Irregularity will be increased if planting is in rows diagonal to the slope and main directions of view.

Tree shelters leaning at various angles produce visual confusion, as well as giving an impression of incompetence and dereliction. Even a single leaning tube catches the eye. They should always be erected vertically, secured by a stout stake. In close views near paths, etc., stakes should be hidden within the shelter to keep the form simple and less artificial.

Fig. 11.18. Brown tubes blend well with a variety of vegetation colours.

Fig. 11.19. Olive tree shelters blend satisfactorily with grass colours.

Fig. 11.20. This mass of tree shelters appears too large for this small scale landscape. Irregular areas of olive green shelters would have reduced the intrusive scale.

Fig. 11.21. Groups of brown and olive shelters reduce the visual impact.

Decaying tree shelters produce unsightly litter, breaking down in strips as the plastic degrades through the action of light and weather. Litter may be hidden in heavy vegetation on restocking sites, but may be more obvious in winter, and on afforested sites and areas dominated by grass and heather. Residual stakes and decayed plastic should be cleared away when the shelters are no longer required.

Felling and restocking

- A comprehensive landscape plan should be made well before felling is due to begin.

- Felling gives opportunities for improvements to the external woodland margins, internal spaces, and species pattern.

- A complete pattern of shaped felling coupes should be designed which appears satisfactory at every stage of felling, as part of an overall plan for harvesting and restocking.

- Felling coupes should be asymmetrically and irregularly shaped to follow landform, and should reflect the scale of the landscape.

- The pattern of open spaces and restocked species should follow that of the felling coupes.

- Felling coupes should be designed for efficient harvesting, as well as appearance, and should not be isolated unnecessarily from road access.

- Visual impact of dead branches in the edge of the coupe can be reduced where necessary by pruning and thinning.

- Scattered whippy trees should not be left standing on felled areas. Irregular groups of well-formed trees near the edge of the coupe are highly desirable.

- Where possible some stands should be retained beyond their economic rotation age.

Planning the felling landscape

Felling brings rapid landscape changes which will last for many years. Careful design to high standards is, therefore, needed. Comprehensive landscape plans should be devised which improve visual quality in the long term. External margins, species shapes, open spaces, and the environment for wildlife and recreation can be enhanced, and intrusive scale and shapes eliminated.

Advance planning of successive fellings, and the landscape improvements incorporated in such plans, are essential for the orderly supply of timber and efficient use of labour and machinery. Any areas of forest where felling should be advanced or delayed for landscape reasons must be identified early and taken into account in marketing proposals.

Felling can reveal fine details of the landscape which may have been hidden previously. Water bodies, open spaces, crags, views from roads, and recreation areas and other significant features should be noted and, where possible, revealed by felling, and the restocking designed so that these features are not hidden by the next crop of trees.

The risk of windthrow may inhibit the felling of irregularly shaped coupes. An assessment of the windthrow hazard should be made, and calculated risks taken. Windfirm features and edges should be included in the landscape design, but restocking should not be carried out to straight edges even if they are windfirm.

The permanent landscape framework

Man-made forest landscapes become more dynamic as successive areas are felled and restocked, and these changes should be balanced by small, but significant areas which are left relatively untouched over a long period. Within these areas mature, high quality, small-scale landscapes can be created where the visual impact of felling is less obvious to visitors, and where open spaces, broadleaves, mature, and dead trees will favour wildlife. The open spaces in such areas are useful for deer management, particularly in forests where there is a low demand for recreation.

These areas can partially link natural features and large open spaces; they should not appear as continuous belts running through the forest, so that areas under more extensive management appear completely separated from one another. The coarse textured, small scale appearance of these areas should not extend onto skylines. These permanent features are often best located close to watercourses, where they follow landform and the small scale is more appropriate in longer views.

Although details of stands and open space can be planned at a late stage, the overall extent of these semi-permanent landscape features affects the whole forest. They should, therefore, be designed in the context of any revised external margins and before the design of individual felling coupes.

Shape of felling coupes

Felling coupes should be designed in the same way as other forest shapes, as outlined in Chapter 6. Coupe shapes are second only to external margins in their visual impact, because of the combined effect of tree heights, shadows, and colour contrasts. They should be irregular and diagonal to the contour, and should reflect the shape of the ground by rising uphill in hollows and extending downhill on convex slopes. The extent of these inflexions should increase with the size and prominence of the hollow or convexity. Coupe shapes should reflect the character of the landscape, being more angular in rugged terrain and smoother on rounded land-form. The vertical height of surrounding stands usually means that intricate shapes in the coupe edge appear 'fussy' or simply cannot be seen, and the fine details of vegetation boundaries are similarly difficult to translate into the edge of a felling coupe. Occasional short lengths of straight edge maybe acceptable if they are diagonal to the contour and do not form right angles. Where a geometric felling shape is unavoidable, make sure that restocking is done in irregular shapes.

Coupes at the forest edge can look like extensions of adjacent open ground and this effect on the overall shape of the woodland should be allowed for in the design.

Felling shapes should interlock with each other and with the surrounding landscape. Belts of trees should not be left, especially at the forest edge. If a felling coupe must be screened, retained

Fig. 12.1. Clear felling demonstrating all the features which cause intrusive shape; long straight margins aligned along contours and vertical to them, Symmetrically placed, and meeting at right angles.

Fig. 12.2. (*a*) An attempt has been made to vary a geometric shape, but the effect is still far too regular. (*b*) Analysis showing numerous points of symmetry. (*c*) Possible alternative coupe layout with irregular asymmetric shape.

(*a*)

Coupes and edge indentations start from the same very horizontal line

Overall trapezoidal shape of coupe is very symmetrical

Division between two coupes is nearly equidistant from the side of each. It is also symmetrically shaped

Indentations of coupe and protrusions of forest are all same size and symmetrical shape

Both felling coupes are too nearly the same size

(*b*)

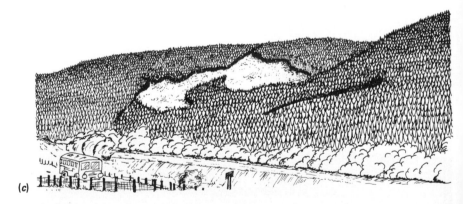

(*c*)

groups should be used. Modern harvesting machines can reach the upper edge of even the steepest slopes, and narrow belts should not be left on such sites, where they look particularly unsightly. If it is uneconomic to fell them because they contain little utilizable timber, some mature areas of suitable shape and scale should be

Fig. 12.3. An irregular felling coupe in a large scale landscape. The shape is generally asymmetric and the detailed indentation of the margin reflects to rugged terrain as well.

Fig. 12.4. A small scale coupe in a more intimate view. The felled area and retentions all appear in scale, and margins fall downhill on convex slopes and rise on hollow ground towards the low point on the skyline.

Fig. 12.5. The interlocking shapes of the clear felled area and retentions have produced a unified landscape. The qualities of this well balanced asymmetric shape, its scale, and the general sense of enclosure offset the conflict of the upper edge with landform, i.e. extending downwards in the hollow.

retained below until all can be felled profitably. Isolated spar, anchor, and support trees should be removed in the final stage of logging, not retained.

Scale of felling coupes

The increasing scale of the landscape on higher slopes and near skylines should be reflected in the size of felling coupes, while smaller coupes are appropriate in shorter views and on lower slopes. Although large coupes need not be unsightly, their scale

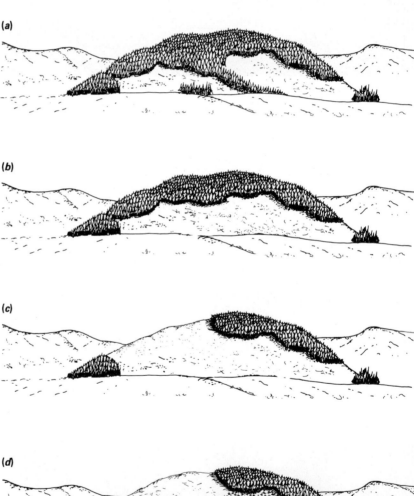

(*a*)

(*b*)

(*c*)

Fig. 12.6. (*a*) Substantial areas can be felled initially, provided scale is controlled by retentions. (*b*) If the retention is lost the scale becomes too large. (*c*) If subsequent fellings are made before the first coupes *appear* restocked, the areas combine into an unsightly large-scale coupe. (*d*) Once tree foliage appears continuous on the previous coupes, felling can take place at an unobtrusive scale.

(*d*)

becomes excessive beyond a certain point. This often occurs where a succession of adjoining areas are felled and before the restocked crop becomes established and 'greens over'.

Conversely, felled areas which are too small and scattered can look 'spotty' and a similar effect can occur with group or shelterwood felling.

Skylines and ridges are always sensitive and should be treated on a large scale, as we saw in Chapter 7. Either a strong forest cover should be maintained, or they should be cleared sufficiently to reveal the shape and scale of the landform. Hilltop broadleaved

Fig. 12.7. The scale of felling is too small, especially at the right, and will create an undesirable spotty effect if continued. These small spaces would look better if amalgamated into an irregular shape by felling the areas between them.

Fig. 12.8. A small felling coupe well situated on the lower slope, running diagonally up a slight hollow, leaves sufficient space for subsequent large scale fellings further up the slope.

Fig. 12.9. A small felling coupe running diagonally up from the base of the slope maintains a sense of enclosure for walkers. The ground flora has survived and logging residue from the previous larch crop is relatively unobtrusive.

woodlands are characteristic of some parts of the south of England and any changes require great care. Narrow belts on the skyline look out of place, especially if they can be seen through.

The apparent scale of felling coupes can be reduced by adopting a very irregular shape, by retaining trees within and around the coupe, and by progressing felling generally downhill.

A highly irregular shape can appear to be divided into smaller parts if some areas are partly enclosed by forest. Retention of trees at the near (usually lower) edge or within the coupe can screen parts of it. Such retentions are more effective if located on eminences in the ground, and should be sited so that the scale of the coupe is not reduced to a narrow sliver. It will be more economic if areas of slower growing crop are retained for this purpose.

Where felling is phased over a number of years, the apparent scale of the coupe can be reduced in views from lower ground by felling progressively downhill from an upper- or mid-slope coupe. Logging residues are screened from close view by a retention which, once the first felling has greened over, can be cleared in two or more successive phases of small coupes running diagonally through it. If they are visible in longer views, substantial areas may also have to be retained on the upper slopes to maintain the impression of a forest landscape.

In narrow valleys one felling coupe on a spur may appear continuous with another on the opposite side, especially in focal views along the valley. We have seen how shape and scale can appear wrong as a result, so it is important that sketches are prepared to

(a)

(b)

Fig. 12.10. (a) Although the coupe is in scale with this small landscape, the retention at the skyline looks too thin and will seem even worse when subsequently felled. The shape is also too square. (b) Additional retention on the convex upper slopes; a possible alternative would be additional felling at the sides to improve the shape.

Fig. 12.11. The apparent size of a felling coupe can be reduced by adopting an irregular shape in which some parts appear to be partly enclosed by forest.

Fig. 12.12. (a) The scale of a large coupe reduced by areas of trees retained within it. (b) If these retentions were cleared, the scale of the coupe would increase dramatically and the woodland on the skyline would look too small.

(a)

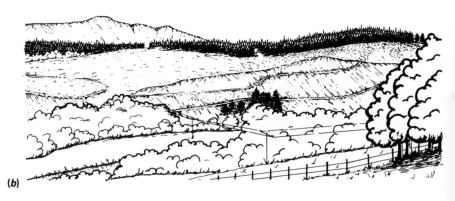

(b)

(a)

Fig. 12.13. (a) Although trees retained at the lower edge have reduced the apparent scale of felling, retained groups (b) would have produced a more unified landscape composition than a continuous belt. The diffuse fringe of trees at the top of the slope is out of scale with the skyline and a more substantial area should have been retained (c).

(b)

(c)

Fig. 12.14. Diffuse belts rarely look satisfactory; solid groups placed irregularly produce a more natural composition while reducing the apparent scale of the felling.

show all stages of felling in sensitive views and that felling is timed to avoid adverse effects.

Efficient harvesting

Felling coupes can and should be designed to be harvested efficiently and in harmony with the landscape. It is essential that timber should be extracted readily to roadside from each coupe, and the requirements of local logging methods in relation to the terrain must be agreed at an early stage of project planning. Their effect on design proposals for all coupes should be assessed, bearing in mind the age of adjacent areas at the time each coupe is felled. Key points are:

(1) every felling should have sufficient road or other access for efficient timber extraction, stacking, and removal.

(2) later fellings should not be separated from the road by earlier restocking or retained stands unless unobtrusive extraction routes through the latter have been identified;

(3) it is often dangerous to work harvesting machines across slopes, safe methods and routes must be selected;

(4) circumstances of road density and terrain difficulty vary greatly, and the best solution must be found for each individual area.

Cases will occur where it is impossible to avoid a retention being isolated from the forest road. If windthrow risk is low, consider

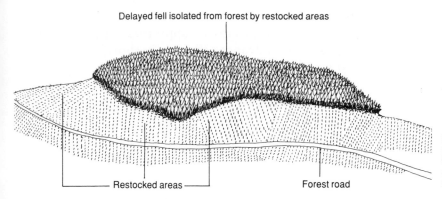

Delayed fell isolated from forest by restocked areas

Restocked areas

Forest road

Fig. 12.15. All felling coupes should be planned with direct access to a forest road; this is not a problem if unobtrusive extraction routes can be found through restocked areas.

thinning the stand heavily before the last adjoining coupe is restocked, so anticipating a substantial portion of the final yield, and carrying out the final felling when the earliest adjoining restocked area is first thinned. This may be a more economic course of action than additional road construction.

Coupe edges

Trees on edges revealed by felling often carry dead branches for about two-thirds of their height, an effect which is most frequent and prominent in spruce. The light brown-grey colour is unattractive, particularly in medium and short views, next to public roads, footpaths, and recreation areas. The 'brown edge' can be ameliorated by a combination of thinning and branch removal, as outlined in Chapter 8.

Even where dead branches are not conspicuous, heavy thinning of coupe edges is beneficial in sensitive landscapes, provided the sites are windfirm.

Logging residues

Measures may be necessary in sensitive areas to lessen the visual impact of logging slash, though in Britain the volumes of residues left on felled areas are much less than in parts of the USA and Canada. In distant views the intrusive light grey colour of dead wood can be reduced by ensuring that the shape and scale of the felling blend well with the landscape. This is especially so where similar colours occur naturally elsewhere, as boulders, rock outcrops and crags.

Impact is greater in shorter views, where the apparent disorder of slash is visible. Its scale can be reduced by screening with retained groups, or by more permanent foreground trees. This is difficult in areas subject to windthrow, and the ideal is to foresee the problem and establish groups of appropriate broadleaves at roadside verges well in advance. If time is short, fast-growing species or even standard trees might be used. Keep them close to the road where they will have the most useful screening effect, on knolls, eminences, and the outside of bends. Visual impact of slash on walking routes can be lessened by:

(1). having footpaths cross and re-cross coupe boundaries, and vice versa, so that walkers only see the felled area for short periods;

(2) diverting footpaths away from felling sites (often desirable on safety grounds) until restocking is established;

(3) locating walking routes within the semi-permanent framework of open spaces and tree groups managed on long rotations, or among lightly-branched larch, pines, and broadleaves;

(4) maintaining drifts of broadleaves and shrubs beside walking routes;

(5) removing slash from areas closest to the path, by chopping, raking, or burning.

Slash may be distributed over the felling area in strips, either as a consequence of mechanical delimbing or deliberately to allow machines to work on soft ground. These strips can be unsightly. Where they run across the slope, the slash often becomes redistributed during extraction and is less noticeable. Strips running up and down the slope are much more common, as this is the direction in which machines work with greatest safety, and such strips are more persistent and intrusive.

In sensitive landscapes it may be necessary to spread slash or burn it. Chopping with brushcutters still leaves noticeable strips, though it may make restocking easier. Planting the slash-free zones and returning some years later to plant in the slash covered strips when the residues have decayed is not satisfactory, and will perpetuate the striped appearance over a longer period.

Visual benefit can be gained by creating irregularity through slight variations in the width, spacing, length and direction of slash strips and reducing or eliminating the parallel alignment of adjoining strips. The extent to which this can be done depends on whether logging methods can be adjusted accordingly and how far any extra costs are justified by the sensitivity of the landscape.

Retention of individual trees after felling

Trees retained on felled areas benefit wildlife and landscape, depending on the increase in diversity, and on their scale, distribution, and individual qualities. Unproductive trees with well-formed crowns are candidates for retention if they form coherent groups or are located close to the edge of the coupe. An even scatter of individuals looks out of scale and 'spotty' and should be selectively felled to form distinct well-positioned groups. Groups in the central third of a coupe will appear isolated and symmetrically placed, and should only be retained if they appear linked to

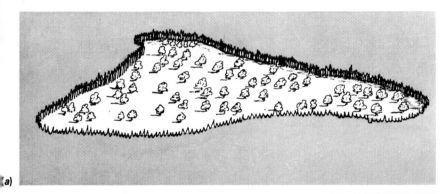

(a)

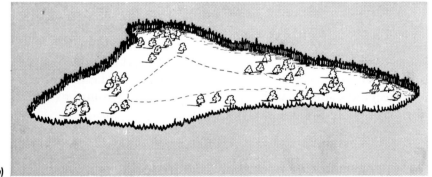

(b)

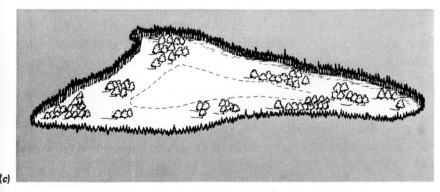

(c)

Fig. 12.16. (a) Individual trees scattered throughout the coupe have a 'spotty' appearance. (b) Discrete groups near the periphery of the coupe improve diversity and scale. (c) (see over) Denser groups improve scale and diversity further. (d) (see p. 240) Groups in the central area of the coupe appear isolated and symmetrically placed (e) (see p. 240) unless linked visually to the edge by other trees.

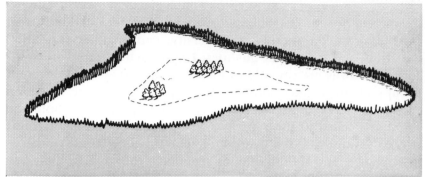

12.16 (d)

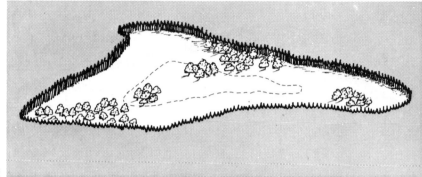

12.16 (e)

the edge by other trees. Leaning, misshapen, or whippy trees with thin or malformed crowns are unsightly and should be felled.

Landscape considerations concerning retained trees will seldom conflict with the needs of wildlife. The latter should only be sacrificed to landscape, or vice versa, after a range of solutions which might satisfy both have been examined.

Design process for felling and restocking

The design of felling is more complicated than for new planting, with many factors to be considered. It is best to design the broader scale first, in longer, more extensive views of the forest. After landscape survey and analysis a complete pattern of felling coupe shapes is devised and an approximate date allotted for the felling of each coupe. The location of contrasting species, e.g. larch and broadleaves, is then decided, followed by any necessary improvements to external margins. If changes to the outside margins are extensive, they should be planned first of all so that the pattern of felling areas fits the overall shape in the longer term. Areas requiring more detailed treatment, such as watersides, recreation areas, or roadside landscapes, should be designed in outline at an early stage if they have significant impact. The details can be designed later.

Accurate sketches should show:

(1) the landscape appraisal and factors affecting design;

(2) pattern of felling shapes and long-term retentions, with proposed felling dates; these shapes should be mapped;

(3) the appearance of the forest at successive stages of felling;

(4) the pattern of species once replanting is complete; map these as well;

(5) adjusted external margins and shapes of internal spaces, areas of recreation, and habitat improvement, and special roadside design.

Landscape appraisal

Trees hide much of the fine detail beneath the canopy, so before doing landscape surveys on the ground it is well worth looking at maps and aerial photographs, and gathering information from local foresters and others who know the area intimately. Local libraries and archives may have useful information.

The landscape appraisal lists the factors affecting the overall design; these vary from one locality to another. Photographs help to distinguish what is clearly visible and what scale of feature might be revealed after felling. Too many details at this stage can inhibit design, so information should be organized according to importance or the ease with which features can be incorporated in the design.

Mapped information should be summarized in a similar way. For example, it may be important to identify soils which will grow larch and broadleaves, but not the full range of soils for evergreen conifers. By transferring *relevant* information on soils, exposure, and windthrow hazard to sieve maps, all these factors can be summarized effectively and then transferred to sketches as a basis for design.

Extensive improvements to external margins should be designed first because the external shape of the woodland has the most dominant visual effect. In the long term, all the adjoining shapes will then appear to be well unified parts of the whole landscape.

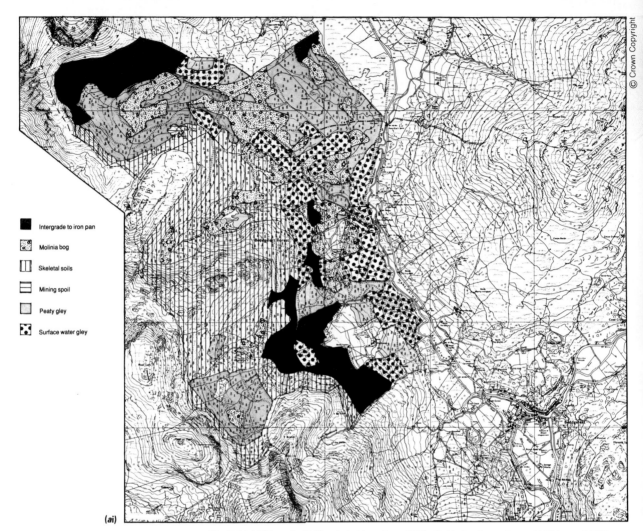

© Crown Copyright

Intergrade to iron pan

Molinia bog

Skeletal soils

Mining spoil

Peaty gley

Surface water gley

(ai)

Fig. 12.17. (a) Maps of soils (i) and windthrow hazard (ii) for Beddgelert Forest, Gwynedd. (b) (see p. 244) Visual forces in landform. (c) (see p. 245) Elements of diversity, viewpoints and visual problems. (d) (see p. 246) Panoramic view of the forest. (e) (see p. 246) Landscape appraisal sketch. Aspects of landscape appraisal will vary from one view to another. (f) Adjustments to the external margin so that it follows landform in a more natural way. The areas shown as broadleaved trees are areas managed at a smaller scale for wildlife conservation and recreation.

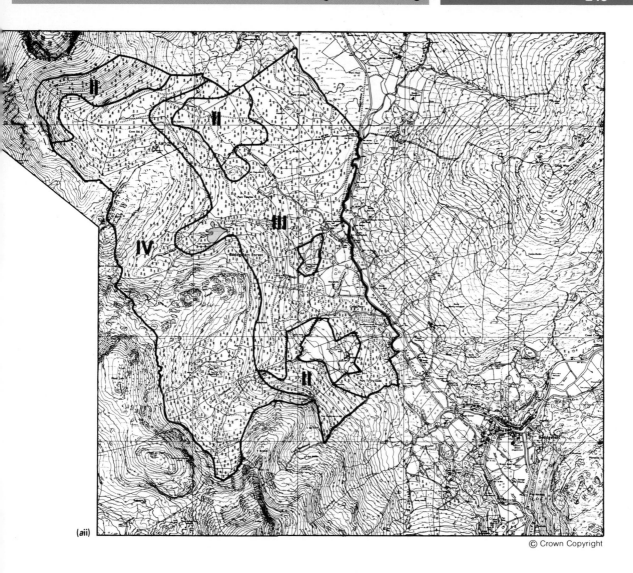

(aii)

© Crown Copyright

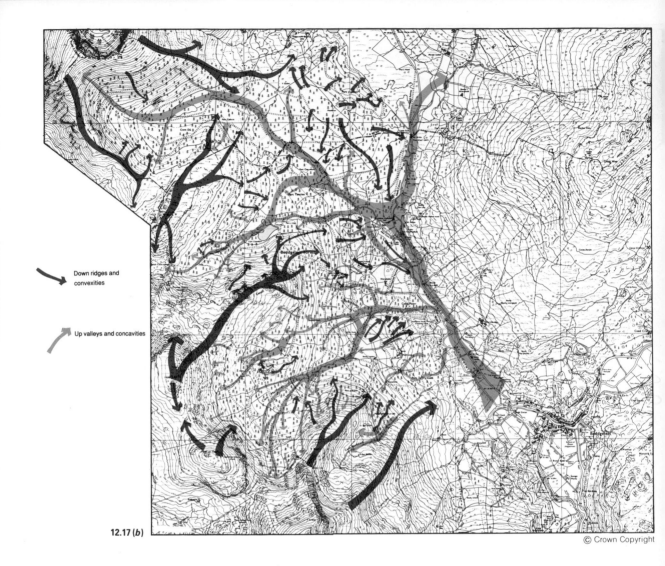

Down ridges and
convexities

Up valleys and concavities

12.17 (*b*)

© Crown Copyright

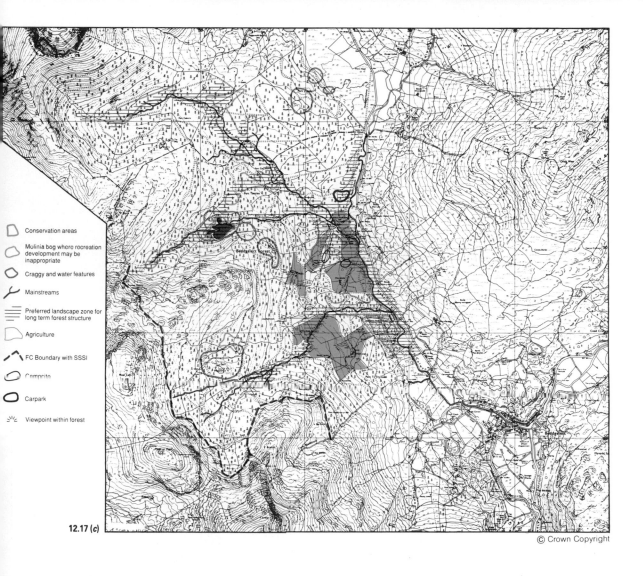

Conservation areas

Molinia bog where recreation development may be inappropriate

Craggy and water features

Mainstreams

Preferred landscape zone for long term forest structure

Agriculture

FC Boundary with SSSI

Campsite

Carpark

Viewpoint within forest

12.17 (c)

© Crown Copyright

12.17 (d)

MAIN LANDFORM FORCES

 Down ridges and Spurs

 Up hollows and convexities

VISUAL PROBLEMS

 Poorly shaped external boundaries geometric or not responding to landform forces

 Areas of differential grazing confusing the external edge of the forest with poor shape

ELEMENTS OF DIVERSITY TO BE DEVELOPED

 Crags

 Waterfall

 Areas which will grow YC8 larch or better

Areas which will grow YC4–6 larch or better

Broadleaves

12.17 (e)

12.17 (*f*)

Windthrow

The effects of windthrow must be considered before planning successive felling coupes. Windthrow is often perceived as a major constraint on phased felling, but some calculated risks have to be taken in the light of realistic assessment of windthrow hazard and previous thinning treatments. Windthrow hazard indices give an indication of critical height, i.e. the height of the trees at which the onset of windthrow can be expected. In areas of high windthrow hazard, critical height is lower and windthrow can be expected to occur at an earlier age; it is, therefore, essential that landscape planning is done early.

Even where the risk of windthrow is very high, any existing geometric shapes must not be repeated in restocking. If necessary this should be delayed in some areas to ensure that irregular forest shapes are developed. Windthrow clearance should not be extended into standing trees in an attempt to 'tidy up' areas to a neat shape. Economic appraisals usually show that it is best to keep standing trees as long as possible rather than fell prematurely, and felling should only take place to windfirm boundaries where these are sympathetically shaped to landform.

Where blown areas look too large their scale can be reduced by planting contrasting species, by leaving some open spaces or by

delayed restocking on part of the area. The shape and scale of all these measures should be carefully designed, and a permanent landscape structure of broadleaves and open space devised to provide attractive windfirm edges for the future. Such edges and species margins should be located on more stable ground as far as possible.

Where windthrow hazard is moderate the landscape appraisal should identify windfirm areas and species which can be incorporated into the design as retentions. Visually intrusive edges should still be avoided.

Pattern of felling coupes

A complete pattern of felling coupes should be designed next, reflecting the scale of the landscape and following landform. It is essential to design felling shapes on perspective sketches and reconcile these with a map. If sketches are based on photographs of existing woodland, shapes should be drawn at canopy level, not ground level, to avoid confusion. This is the most critical part of design and it is worth spending as much time on this stage as on all the rest together. Poor shapes can seldom be improved by other aspects of design, and are costly to adjust later. The designer should only move on to the next stage when no further improvements seem possible.

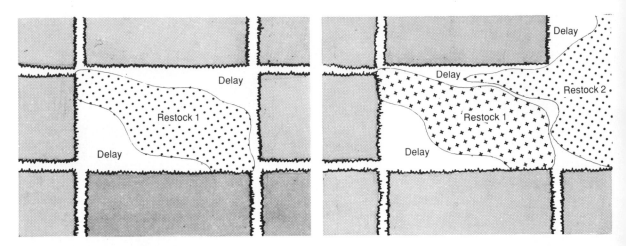

Fig. 12.18. (above) How partially delayed restocking can be used to adjust geometric shapes. This approach should only be used where it is impossible to undertake irregular felling.

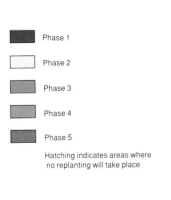

Phase 1

Phase 2

Phase 3

Phase 4

Phase 5

Hatching indicates areas where no replanting will take place

Fig. 12.19. Beddgelert Forest. Felling coupe shapes and improved external margins.

The shape of both early and later coupes will be affected by their common boundary. It is often better to design the more prominent areas first, even if they are to be felled later. Make sure that harvesting can be carried out safely and efficiently, and that each coupe has adequate access for timber extraction.

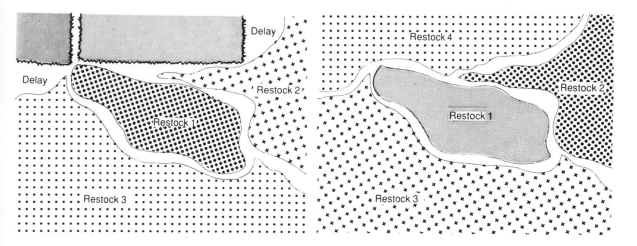

Timing and order of felling

When a mainly even-aged, first rotation, planted forest reaches felling age, the opportunity to increase diversity is not to be missed. This can be achieved by extending the range of ages, so bringing contrasts of stand texture and height. Felling and restocking in most landscape compositions should be phased over a number of years to create a pattern of mature stands, felled areas, and young crops, in scale with their surroundings. Where windthrow or marketing commitments prevent the retention of mature stands, large areas which have been felled at one time may have to be replanted in distinct stages, with 7–10 years interval between each stage.

Once a complete pattern of satisfactory shapes has been established, the order and timing of each felling can be decided. These are influenced by the need for felling to take place as close as possible to the age of maximum economic return, at the right scale, and the rate at which restocked areas 'green over'.

Final felling is the first major landscape change since planting. It

Fig. 12.20. Comprehensive felling and restocking design. (*a*) At Coed y Brenin in Wales felling has been going on for some time and has now come to be planned in a particularly sensitive area. (*b*) (see p. 254) The visual appraisal. The convex landform predominates with a major concavity leading from the river valley up the focus of the view. An area of large Douglas fir is to be retained in the sheltered ground in the valley bottom. (*c*) (see p. 256) The felling is planned for the whole area with coupe shapes related to landform and the scale increasing from valley bottom to hill top. The felling will occur at 7-year intervals to ensure a high degree of diversity. (*d*) (see p. 258) When the felled areas are replanted larch will be used to highlight the convex landform and broadleaves used to connect the forest to the surrounding landscape. (*e*) (see p. 260) The forest after the first phase is felled. (*f*) (see p. 262) The forest after the last phase is felled, and the previous phases have been replanted and are growing at different ages producing different textures.

(*a*)

should be started on a modest scale, preferably in less visible areas, and so introducing people gradually to the series of changes which will take place in future years. The landscape plan should be prepared early enough to allow a few small areas to be felled prematurely if necessary. The most prominent parts of the landscape, such as ridges and skylines, should be felled after a number of coupes have been restocked.

The time taken to achieve a green appearance (shortly before canopy closure) depends on how quickly young trees can be

established. It can vary from 8 to 15 years, depending on site conditions. Efficient restocking is specially important in the early stages of felling, so that the cutting of adjoining areas is not delayed unduly. It is possible to reduce the interval between felling adjoining areas by planting at closer spacing at the edge of the restocked area; this gives a defined green shape when the interior appears only diffusely covered with trees.

To maintain the scale of the woodland landscape and obtain 'significant contrasts in texture, each cycle of felling should prefer-

Skirt of mature conifers
Public road (minor) runs through

12.20 (*b*)

ably last as long as it takes restocked areas to 'green over' and each coupe should be assigned to a particular period on this basis. It is usually possible to devise the landscape plan so that an individual coupe can be felled at any time within a period of 6–10 years. This gives some flexibility for timber marketing and management of labour. Where a number of different coupes are to be felled in a single period, they should vary in size and be irregularly spaced and positioned in the overall composition.

Using a variety of species in restocking increases diversity, parti-

orest walks and viewpoints
vell used in this part

Rough textured open space
gives some diversity

cularly in later years when differences of age of crop are less obvious.

Having allotted each coupe to felling period, it is essential to illustrate the appearance of the forest at future periods. Illustrations showing the next three or four periods are usually sufficient. This ensures that an asymmetric balance is maintained and that the scale does not become too fragmented. Illustrations should

Phase 1

Phase 1 (do not restock)

Phase 2

Phase 3

Phase 4

Long term retention

12.20 (c)

take tree height into account, showing different cycles of restocking in different colour shades, as described in Chapter 16, Design techniques.

If the illustrations point to visual problems in years ahead which can only be solved by adjusting coupe shapes, revisions should be made and illustrated. If the shapes have been well designed a major revision should not be necessary, but it has to be understood

that serious design mistakes must be rectified at this stage. Shape design has an effect on the future landscape which cannot be effectively ameliorated by any other means.

Felling design in flat landscapes

On flat ground the vertical edge of the woodland dominates minor variations in landform almost completely. Although variations

12.20 (d)

in height are important, the shape of the woodland edge still affects perception of the mass of the forest and the open spaces within it. The overall impression of the forest is gained almost entirely from the sequence of views from roads, with few points from which an overall view can be obtained. Forest design should, therefore, aim to develop an interesting succession of small, permanent roadside spaces, supplemented by occasional views of well-designed felling coupes. The quality of the landscape depends far more on the interest created by the designer than on avoiding

intrusive mistakes. It will only be successful if both designer and forest managers develop the landscape in a whole-hearted manner, rather than doing the minimum necessary.

Views from the road should be planned as outlined in Chapter 13. Interesting features are often scarce, and to counterbalance this uniformity, the quality of the various spaces within the forest must be higher than in more diverse landscapes, making the most of any features present.

In addition to contrasts of open spaces, the landscape of flat

12.20 (*e*)

forests can be varied by differences in tree age and height. Contrasting species are most effective in groups within a felling coupe or along an edge, as are any existing groups of broadleaves revealed by felling. As in other forests, a semi-permanent framework of retained drifts of mature trees, open spaces, and diverse edges should be planned along footpaths and other routes used by visitors. Such features should not be allowed to create intrusive small scale effects on distant edges or any prominent skylines.

Design of felling coupes

Felling coupes should be designed as part of the roadside land-scape, so that they take in both sides of the road and the space appears to flow from one side to the other. This effect is obtained by placing them slightly offset on either side of the road, with views angled so that motorists see them obliquely.

In general, the scale of felling coupes should be small enough to maintain a dominant impression of a forest interior. When a woodland edge of tall trees is seen from more than about 300 m

12.20 (f)

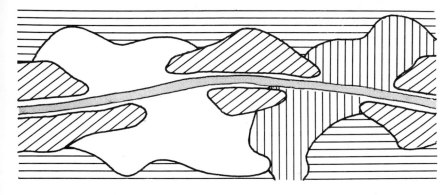

Fig. 12.21. Coupes making gaps of at least 80 m at the roadside to allow the flow of space from one side of the road to another.

this impression is lost, so coupe edges are best kept mainly within that distance of roadside.

The plan shape of coupe edges follows the principles of design of internal space, as set out in Chapter 9. Minor undulations in the ground occur in even the flattest landscapes and on slopes steeper than about 3 per cent of coupes can be shaped to follow landform. Early fellings should if possible be made on lower ground, so that the height of the remaining trees emphasizes rather than negates the landform. Views from the road are better when they are across slightly concave ground than over convexities.

In plan, coupe edges should be curved and asymmetric. Steps in the edge of the coupe add interest where required, and small retained clumps provide focal points. They must be carefully located

Fig. 12.23. An oblique view opened up through a felling coupe in Aldewood Forest, Suffolk. The eye is drawn effectively through the space, giving a sense of distance, but the far edge is so distant that the impression of a forested landscape is lost.

Fig. 12.24. The concave shape of the ground allows the sinuous lines of this coupe to be readily appreciated, giving it a high aesthetic quality. The variation in tree height adds diversity and the retained clump provides an interesting accent.

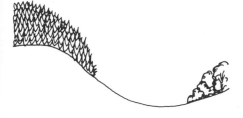
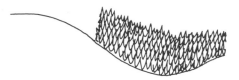

Fig. 12.22. (*a*) Topography emphasized.   (*b*) Topography negated.

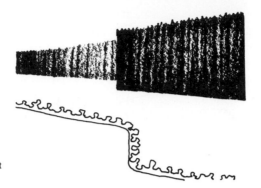

Fig. 12.25. A 'step' in plan giving an accent in perspective.

to maintain an asymmetric balance, along with access for harvesting once the surrounding crop has grown up.

Variations in height

Where one species is dominant, variations in age of trees are important for diversity, and the resulting differences in height are most obvious in the edges of the forest and felling coupes. In larger forests and woodlands the long-term aim should be to have three widely different age groups, if significant height differences are to be continuously visible, i.e.

(1) low   felled and newly restocked areas;
(2) medium   young trees, at least 15 years old;
(3) high   tall trees, mostly within 15 years of felling age.

Fig. 12.26. A felling coupe of irregular shape in a flat broadleaved forest with retentions of varied height around it. Salcey Forest, Northamptonshire.

Since it would involve financial sacrifice to change directly to such a grouping from an even-aged crop, the change is best made gradually. Felling can be spread over a period of up to 20 years, with a proportion of premature and delayed felling. Windthrow can, on occasion, provide an element of unplanned premature felling which helps to widen the age–class spread. The penalties of delayed felling can be reduced by heavy thinning of the longer retentions at the age of maximum economic return, always provided the site is windfirm. A general impression of forest cover can be maintained in a retention if half the remaining trees are removed after a normal thinning. A greater tree density is needed as slope increases, in small stands, on skylines, and close to roads and paths where the crowns appear to overlap less.

Fellings at roadsides can achieve a satisfactory age-class range, with a minimum age difference of 10 years between adjoining areas by felling a sequence of coupes at economic rotation −15, −10, −5, 0, +5, +10, +15 years. These different ages should be planned to achieve an interlocking pattern of various blocks of woodland. Avoid having different ages meeting at corners.

A model of the proposed design of the forest is useful in co-ordinating the various age classes. The appearance of felling coupes can be simulated an a 1:500 scale using strips of corrugated cardboard pinned to foam board or cork sheet. The vertical scale of the cardboard strips can be exaggerated by up to 130 per cent without distorting the spacial quality of the felling coupe. The model can be viewed by model scope or from an open edge.

If delayed felling has to be brought forward, perhaps because of

Fig. 12.27. Different heights of forest stands should be planned in an interlocking pattern. Overlapping masses of different aged stands: (a) well interlocked; (b) not interlocked; (c) a coupe design which makes use of interlocking stands.

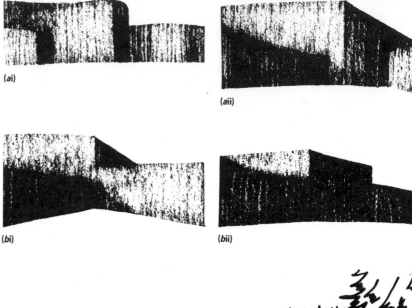

(ai)

(aii)

(bi)

(bii)

(c)

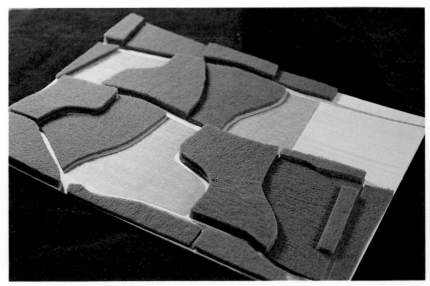

Fig. 12.28. Model prepared in layers of foam sheeting to illustrate principles of design for different ages of Thetford Forest, Suffolk.

disease or insect damage, a predominantly forested impression and the scale of the coupes can both be maintained by means of smaller retentions of irregular shape around the edge of each space. They should appear as overlapping groups, not as continuous belts, and

Fig. 12.29. Model of felling coupe using cardboard strips on foam board to demonstrate internal perspective; note overlapping layers of tree canopy.

Fig. 12.30. Overlapping layers of different stands drawing the eye into the distance. New Forest.

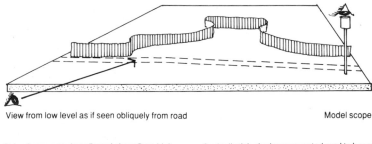

View from low level as if seen obliquely from road   Model scope

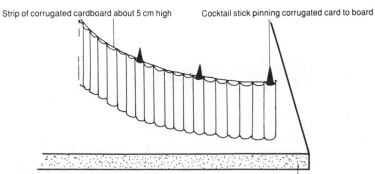

Strip of corrugated cardboard about 5 cm high   Cocktail stick pinning corrugated card to board

Plan of coupe shape at 1:500   Foam board

Fig. 12.31. Modelling technique for felling coupes on flat ground.

Fig. 12.32. A large roadside felling reveals stands of varied height (centre) which are insufficient to provide a focus in this scale of space. The position of the foreground opening means that this feature is not seen in the oblique view which a motorist would see.

their shape should reflect any visible landform and be strongly curved in plan. These areas might also be underplanted with contrasting species (e.g. broadleaves in a pine forest), so as to give an impression of stronger enclosure and greater diversity.

Edges

The variety of edge treatments described in Chapter 8 can be used in forests on flat ground. A sufficiently bold scale is needed along edges seen at greater distance or viewed from moving cars. Retentions at the edge should be large enough to appear as solid mass and should be irregularly spaced and varied in size. This is specially true for back-lit edges, i.e. those facing between north-west and north-east, which often appear deeply shadowed. Broadleaved groups in these locations should be kept well clear of tall conifer edges so that their contrasting foliage is not hidden in shadow.

Coupe edges can be diversified in similar ways or thinned to varying degrees. Avoid straggly groups of trees appearing isolated by doing additional thinning only where the edge of the coupe retreats into the mass of the forest.

Fig. 12.33. A back-lit edge of Corsican pine, scale of which is in marked contrast to that of the surrounding hedgerow. Broadleaved trees in front are hidden in deep shadow.

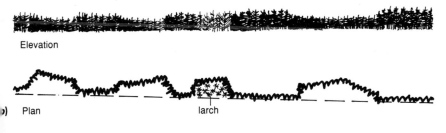

Elevation

Plan   larch

Fig. 12.34. Reducing the impact of a dark conifer edge. (*a*) Varying a long edge of pine by a series of irregular steps in echelon with broadleaved groups kept sufficiently clear of deep shadows. (*b*) A long edge of pine varied with bold indentations to give shadow breaks (one filled with larch for colour contrast). (*c*) Reducing the visual impact of the solid 'wall' of dark conifers by 'off site' planting. This practice is only effective if viewpoints are limited.

Windrowed stumps

In some areas it is the practice to remove stumps after felling, to reduce the risks of further infection by *Heterobasidium annosum*, for example. These stumps are unsightly, particularly when placed in parallel windrows. Although windrowed stumps can be screened from view to some extent, there comes a point when the enclosing forest edge is more intrusive than the stumps, and open views are needed. Unsightly appearance of stumps at these points can be reduced by:

(1) aligning windrows at an oblique angle to the road;

(2) where windrows are at right angles to a road, varying the distance from the end of the row from 0 to 30 m;

(3) where suitable machinery is available the stumps can be collected into low mounds in roadside spaces, placed to frame any oblique views.

It is essential that the restocked edge should extend far enough for the young trees eventually to screen windrows completely from the road.

Fig. 12.35. Gap through windrowed stumps drawing the eye obliquely from the road.

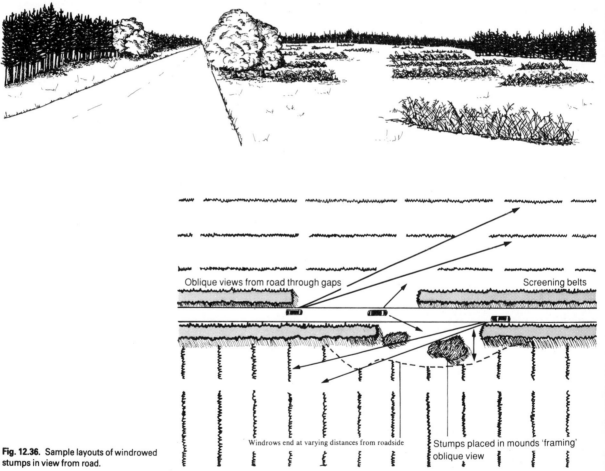

Fig. 12.36. Sample layouts of windrowed stumps in view from road.

Roadsides and recreation

- The movement of travellers along roads through forests and their enjoyment of the adjoining landscape are prime considerations in roadside design.

- An interesting sequence of places should be visible along the road, co-ordinated by a succession of varied roadside spaces.

- Roadside spaces should be planned to dramatize natural features and the motorists' sense of movement, as well as increasing visual diversity.

- Landscape features should be recorded and if possible incorporated at an early stage of design.

- Quality and character of the roadside space can be enhanced by edge treatment and additional planting in the verge.

- Walking routes and rights of way should be an integral part of forest landscape design.

- High quality, small scale landscapes with potential for recreation should be identified and conserved.

This chapter deals primarily with the design principles and processes for the forest, beside public roads and forest drives. The same principles apply to the design of footpaths and bridleways, though here the slower speed of travel and the emphasis on recreation require a greater attention to detail and provision for more viewing and picnicking. The forest landscape beside public roads and paths presents an outstanding opportunity to display the forest as an enjoyable place, tempting passing travellers to stop and visit.

Speed, sequence, and the sensation of motion affect the traveller's perception of the landscape. Scale, in particular, is affected, the broad landscape becoming more dominant at greater speeds while fine detail is only appreciated at a more leisurely pace.

Natural landscape features, spaces, and edges also influence roadside design. Any views of natural features which can be seen from the road are a useful contrast to the woodland, and should be mapped at an early stage of design. These form the basis for a sequence of events which can be co-ordinated and emphasized by a succession of spaces in the forest. The diversity and sense of motion provided by roadside spaces can be increased further by edge treatment.

Much of the approach and practice described here has been developed as a consequence of work published in the USA by the USDA Forest Service and in particular Appleyard, Lynch and Myer (1964).

Principles of roadside design

Roadside design is carried out for the benefit of a captive, inattentive, mobile, and sometimes apprehensive audience whose perception of the landscape is strongly affected by the enclosure and motion of the motor vehicle and by the rate and sequence of landscape changes seen while driving. The effect is rather like a three-dimensional film seen through the frame of the windscreen and windows. The frequency of change and incident seen by the traveller should be about 4–8 seconds, though care is needed to ensure that rapid changes near the road do not confuse and frustrate the enjoyment of an interesting landscape beyond.

The sequence of landscape change also affects the traveller's enjoyment because each sensation is influenced by what precedes it and expectation of what is to come. A wide open view will seem more impressive if suddenly revealed after a strongly enclosed stretch of road than if it appears in fits and starts. The entrances and exits of the public road into and out of the forest are, therefore, of particular importance because the openness of the landscape changes so dramatically. This change often has to be balanced by a sense of continuity, perhaps created by distant features revealed from within the forest which have previously been seen from outside.

The sequence of a roadside landscape should appear interesting to people travelling in either direction. Vehicles join and leave a route at different points, so road junctions should have a strong sense of place and appear as a logical end to sequences of approach views.

The view from vehicle

The perception of landscape changes as soon as one enters a car, bus or other vehicle, even when it is stationary. Large scale landscapes immediately seem smaller, almost as though our ability to cover the distance by movement affects its apparent size. Perceptions of landscape are also affected by position in the car. Drivers tend to concentrate more on the road ahead while rear seat passengers are forced to look to the side. The view of the front seat passenger is probably a reasonable compromise on which to base design. Roadside landscapes should be assessed and planned as far as possible from a car. Planning them on foot often results in too small a scale for the motorist to have time to enjoy.

A driver's view is concentrated close to the road and this area has been estimated to receive about two-thirds of the attention of front seat passengers. Though the remaining third of travelling time may be spent looking at more distant views, front seat passengers rarely look at right angles to the road. Acute angles of view, areas close to the road, the outside of bends, and the ends of straight stretches are, thus, particularly sensitive elements, reinforced by the visual impact of the road itself, pointing away from the motorist towards a vanishing point.

In contrast to the general pattern described above, the occasional periods when more attention is given to the wider view usually occur after visibility has been strongly contained for some time. Following such restrictions open views seem a dramatic contrast. These can be provided by stretches of forest or woodland

Fig. 13.1. Glimpsed views into the forest which are interesting to the walker are often missed completely by the motorist.

Fig. 13.2. The road draws the eye to the gimpse of water through a screen of tree trunks on the outside of a bend.

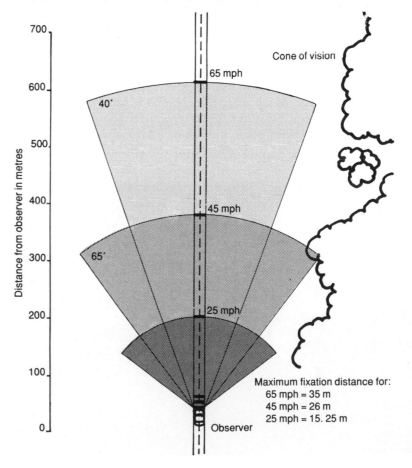

Fig. 13.3 Angles of vision as seen from moving vehicles of various speeds

close to the road, especially amongst more open agricultural landscapes.

As speed increases, motorists' attention becomes concentrated closer to the road and further ahead. The driver's angle of visual attention of about 100° as 25 mph will decrease to less than 40° at 65 mph. Foreground detail is unnoticed and subtle changes appear insignificant. The motorists' attention can be drawn to attractive features such as clumps of trees near the road which appear to move in front of a more distant background, though if there are many such features they become lost in a blur at higher speeds. A few carefully positioned, well maintained clumps are far better than numerous individuals.

The motorists' attention also becomes focused closer to the road at points where the driver has to make decisions or where greater driving skill is required. Road junctions, narrow bridges, sharp bends, steep hills, and blind summits are all places where the landscape should be quite simple. Sight lines necessary for road safety must be kept open, intrusive design avoided and points of interest maintained at a low level in such situations. The landscape should consist of clearly identifiable spaces, preferably quite enclosed, so that no wide view distracts attention from the road.

The sense of motion

The impression of moving across the landscape is a major part of the pleasure of travelling. Changes in landscape character, spaces

Fig. 13.4. Occasional clumps of trees close to the road heighten the sense of motion through the landscape. Very regular roadside avenues can become monotonous and create an annoying flicker at certain speeds.

Fig. 13.5. Clumps of trees close to the side of the road receive detailed attention from motorists. This group of roadside birch would benefit from some pruning to reveal stem colours.

Fig. 13.6. A diverse roadside landscape with enough variety close to the road to enhance the sensation of speed. Many details will be missed by traffic at 60 mph.

opening out and closing in, near objects flashing past occasionally, landmarks seen from different viewpoints, all heighten the sense of motion. The physical movement of the vehicle contributes to this sensation and speed always seems greater on steep descents and sharp curves. The perception of travelling is sharpened by patches of light and shadow on the road, constricted spaces, tree canopy overhead, and occasional features near the verge. A landscape with these features can make 40 mph seem faster than 60 mph on a completely open road.

Roadside landscape design should, therefore, develop constricted

(a)

(b)

Fig. 13.7. Varied spaces and roadside features increase the enjoyment and sense of movement through this part of Thetford Forest, Suffolk. (*a*) Enclosed landscape flowing from one side or the road to the other, round incidental broadleaved trees casting shadows on the road. (*b*) The end of the enclosed space and the entrance to the forest. (*c*) (see p. 280) The strongly constricted space creates a sense of arrival within the forest. (*d*) (see p. 280) The light ahead creates a sense of anticipation just before (*e*) (see p. 281) arriving at the broader space beneath the tree canopy.

spaces to emphasize the physical motion of the vehicle on bends and steep hills. Forest edges should be kept further back on gentler road alignments, with occasional features near the road to prevent the impression of movement being lost.

Appraisal and planning of roadside landscapes

All the factors mentioned which affect the travellers' view of the roadside landscape have to be reconciled with the features of individual sites. A careful appraisal of the landscape identifies

13.7 (c)

13.7 (d)

problems which the design must overcome and opportunities to show features of outstanding quality.

Maps and aerial photographs help to identify features of interest and any which are screened by the forest from the road can be located. It may be necessary to draw topographic sections to determine the extent of forest clearance required to reveal a particular feature.

The sequence of views should be assessed in both directions by a passenger in a vehicle travelling at normal speed. Subsequent appraisal from a vehicle with stops to take notes is often necessary, but points identified when stationary should be verified at motor-

(e)

ing speed. This examination of the landscape will show where awkward changes, uninteresting areas, expectation and anticlimax, contrast and continuity, improve or detract from the pleasure of the journey. The traveller should encounter a series of attractive landscapes at spaced, irregular intervals along the road. A basically good standard of landscape is required everywhere, with outstanding features displayed selectively; too many of these in a quick succession carries a risk that the remainder will appear dull.

The appraisal should start well before the entrance to the forest, even if only a short stretch of road passes through it. The sequence will probably start at a significant settlement, hill pass or road junction, and the impression of the forest will depend on what the traveller has seen earlier. After a generally open approach the constricted space of a forest may seem a striking and welcome contrast, while a partially wooded approach may require a succession of open and enclosed spaces to be continued.

The general character and quality of different stretches should be identified during appraisal. Attractive character can then be maintained and heightened by design, while that of poor landscapes should be improved. The nature of improvements will depend on the need to contrast with or extend adjoining areas in the sequence of changes of view along the road.

Focal points on the road are also identified, such as outstanding views, places where landscape character changes suddenly, or road junctions. The entrance and exit to the forest are especially important and do not always coincide with the boundary of ownership. To the motorist they are where the road passes from the main body of woodland into predominantly open surroundings, or *vice versa*, and should be emphasized by a strongly enclosed space, almost like a gateway.

Eyesores and structures are memorable features which should be identified. An eyesore should be removed if possible, or radically improved, rather than screened. When screening is the only practical option, a well designed fence or dense planting of fast growing species should be used so that it does not detract from the benefits of other roadside improvements for longer than necessary.

The number of man-made structures at roadside should be reduced to an absolute minimum, in order to emphasize the natural qualities of the forest. Fences no longer required should be removed and others sited inconspicuously away from the road. The number of signs and notices should be limited to essentials, using symbols instead of words wherever appropriate. All necessary structures should be designed as simply as possible at a robust scale, and be well constructed and maintained. Those which have to be repeated at intervals along a road, e.g. reflectors to deter deer, should be spaced irregularly.

Distant views are a welcome contrast to the enclosed and canopied forest, and the potential for such views should be assessed on gentler downhill sections of roads, at summits and on the outside of bends. If the same view can be seen from different places, the viewpoint with the best quality should be selected, e.g. over foreground water or in a focal composition. Subsidiary views should also be chosen where they can provide a sense of anticipation.

In steeper topography, good views of waterfalls or cascades can often be obtained above the road where it approaches watercourses. Farms, unplantable land, and other open space within the forest which can be seen from the road should also be recorded for the contrast they provide.

Exciting road alignments, such as steep descents and tight bends, should be recorded with a note that enclosed roadside space may be needed to heighten the traveller's sense of speed. Road summits should also be recorded because a long ascent gives a sense of anticipation to a revealed view. Some screening trees may be necessary at such points so that the view is seen first from the best place, to avoid any anticlimax.

Landform at roadside is an important part of appraisal, as roadside edges should rise in hollows and descend on ridges. The appraisal should make a record of landform on a contour map, supplemented by field inspection or aerial photographs. Where

Fig. 13.8. When distant features are continually visible from the road, key views, e.g. over water or in focal compositions, should be selected for emphasis and framing. Cul Mor, Suilven, and Canisp seen from Benmore, Sutherland.

Fig. 13.9. An attractive view of a loch, hidden from view (*a*) until revealed at the summit (*b*).

(*a*)

(*b*)

contours give insufficient detail and the landform is obscured by forest, a site survey will be necessary for felling design. If felling is carried out in stages, the shapes of clearances can be adjusted as landform is revealed.

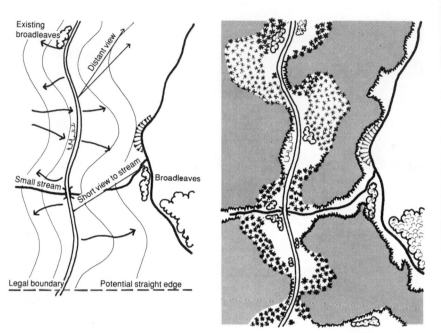

Fig. 13.10. An example of how roadside spaces should relate to landform.

Fig. 13.11. Trees of different ages add diversity to the landscape. A more flowing edge meeting the road at an oblique angle would be better.

The main characteristics of any existing forest stands and road-side edges should be recorded for their present and potential aesthetic qualities, and for the constraints which might be imposed on the design. Broadleaved trees, larch, and pine introduce contrasts of colour and form to evergreen conifer woodland, and light-canopied species allow more light onto the forest floor. Edge diversity is increased by variations in the size of trees and areas of checked growth can sometimes be usefully revealed by a small amount of felling.

The edges of the woodland have a profound effect on the road-side landscape, and areas with unattractive spaces and intrusive edges should be recorded. Edges close to the road are an asset only where they follow a sharply sinuous road alignment; in such places they should be maintained for short distances to increase the sense of movement. Roadside edges which are straight, parallel to the road, equidistant on other side, and too strongly enclosing should be noted for improvement as soon as possible. Opportunities to do so at time of restocking should not be missed; if there are doubts about windfirmness, work should start early to give the edge a better chance to stand. An assessment of windthrow hazard and whether trees are approaching or at critical height will indicate where there may be constraints on design. These are not an excuse for inaction, rather a warning of where a calculated risk may be necessary.

Fig. 13.12. A forest landscape which is prominent in the view from the road. Good design is marred in parts by geometric side boundaries, which need improvement.

Areas of forest which are highly visible from the road should be noted as sites requiring particular care in the planning of future operations.

The detailed appraisal of roadside landscape should be amplified by an assessment, and record of the more general features and character of the wider landscape, made after a final car trip along the road in each direction. General impressions might be that the landscape is too enclosed, too uniform, or that it is sufficiently diverse in detail, but lacking any memorable features or drama. These impressions should guide the broad planning and design.

Development of roadside landscapes

The general statement of problems and features in the landscape appraisal indicates broadly how the series of roadside spaces should be designed. Opportunities for important views suggest suitable locations for open space, while steep twisting road alignments suggest where the forest edge should enclose the road strongly. Steep, forested valleys often feel oppressive and large open spaces may be needed every 2 or 3 miles as a relief from the enclosure of the woodland. Conversely, a strong sense of enclosure may be more important than views where an otherwise open road passes through a short stretch of forest. There may be instances where some less significant views would be better screened, and others given slightly less emphasis, to make the overall sequence appear more interesting.

Fig. 13.13. Views have to be opened fully for the motorist to enjoy the landscape beyond. Although the open space is a welcome contrast to the woodland, such a limited glimpse is only justified as a foretaste of a fully revealed view.

Fig. 13.14. A simple, light background provides an effective contrast to the trunks of the trees.

The character of the exposed open views should be reflected by the forest in the foreground, as described in Chapter 5. Views should be visible at a scale large enough for the speed of the motorist, without leaving major features half hidden by trees and with some allowance for subsequent tree growth. It should be clear what is being shown at each point; a good view over the forest should not be confused by trunks of large trees in the foreground since the two can probably be shown separately to greater effect.

The pattern of roadside spaces will be determined partly by potential views identified in the appraisal. On featureless stretches of road the overall design should be developed as a sequence of roadside landscapes in which any natural character should be enhanced. The roadside space should vary in width and extend first on one side of the road, then on the other. The scale and alignment of the space should reflect the speed of traffic. Detailed guidance on design of internal spaces is given in Chapter 9. Where an additional incident is required to add interest, an open area extending obliquely into the forest can be introduced to provide a view. The space should be varied in width with occasional clumps of trees to give a sense of depth, and simple enough to be enjoyed instantly by the motorist.

Places where a road passes under the forest canopy are highly memorable and invoke the essence of woodland landscape. They occur frequently on narrower roads among broadleaves, but are rarer in upland forests and on the wider-verged main roads. The additional effort of managing large trees close to the road is worth while in certain limited areas at key positions, but it is essential to take precautions to protect public safety.

The form, definition, and proportions of these canopied spaces give an outstanding quality to the landscape if properly designed and managed. Broadleaves and conifers create quite different qualities of space, the former often having the appearance of a vaulted roof while the latter demonstrate soaring verticals.

Fig. 13.16. Exciting abstract rhythms of road alignment, landform, and forest edge which could be improved by one or two carefully placed clumps. More extensive planting of the verge is likely to destroy the sharp contrast of verge and woodland on which this quality depends.

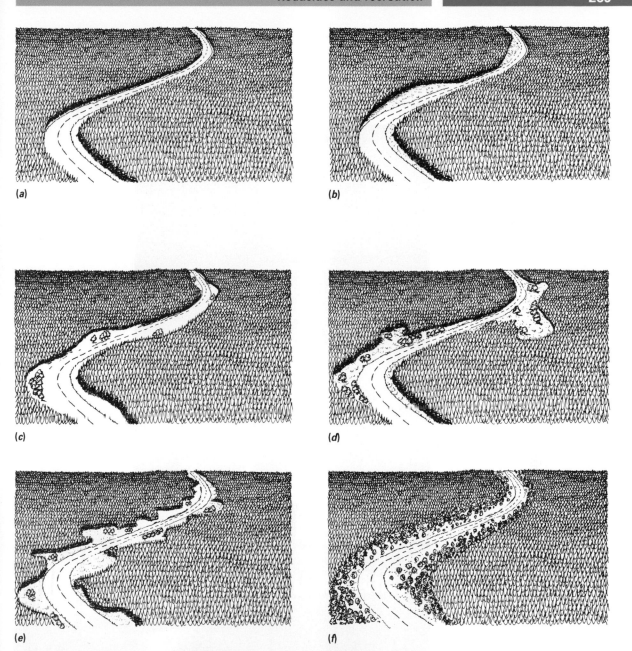

(a)

(b)

(c)

(d)

(e)

(f)

Fig. 13.15. (a) Uniform forest edge on either side of the road appears uninteresting and oppressive, and may disorientate the traveller. (b) More sinuous space, flowing from one side of the road to the other, can create interesting rhythmic shapes; appropriate for fast traffic (60 mph), but still rather too uniform. (c) Greater variation in the roadside space with clumps of trees giving the traveller a greater sense of movement through the forest and providing points of interest (50–60 mph). (d) Additional open space used to create 'false' views into the forest to excite the attention of the traveller. Most suitable where there are few good views to the wider landscape (45–60 mph). (e) Too small a scale of variation in space for such a sweeping alignment and too repetitive. This scale of design is more suitable for forest drives, but would need to be supplemented by greater detail of edge, natural features, and individual trees. (f) The scale of variation has become so small that the overall appearance is highly uniform. Although the texture of the edge is coarser than (a), it is just as monotonous.

Shape of forest edges should follow landform, once the general distribution of open space along the road has been planned. This may involve some compromise between the two, but intrusive shapes cutting across landform should always be avoided. A curving edge, diagonal to contour, is a prudent standard to adopt.

Fig. 13.17. Great size and soaring verticals evoke the characteristic internal landscape of conifer forests. (*a*) Cathedral Grove, B. C., and (*b*) Strathspey, Morayshire (courtesy Oliver Lucas).

(*a*)

(*b*)

Roadside spaces are often visible from the surrounding area, especially where they traverse steep slopes. Although a narrow roadside space can be hidden by tall trees, anything which permits a view out will inevitably be a visible gap. These spaces need not appear intrusive in the wide view if their shape and scale are correct in the relevant view. Design is straightforward on lower slopes and in small scale landscapes, where shapes follow landform in the normal way. In large scale landscapes it is more difficult to design a succession of roadside gaps, especially near the skyline where the visual problems are the same as for other small spaces. Closure and rhythm can be used to give the impression of a larger pattern in some situations, but the needs of the wider landscape may require a greater compromise in the traveller's view from the road.

Views from the road may be revealed from different places as successive areas are felled and restocked, and permanent open space will subsequently look less intrusive if located at the boundary of adjoining coupes. The scale of roadside openings can be introduced into larger scale forest landscapes as part of an overall pattern with other open space, e.g. beside streams and around rock outcrops. Views from the road usually fit the landform quite well since spaces below the road on spurs reveal views to the wider landscape, while streams and other water features are often seen above the road in re-entrants.

The general location of these spaces can be identified as part of the roadside landscape first and then marked on a sketch. The

Fig. 13.18. Beech in the Chilterns demonstrates the more intimate canopied qualities of internal broadleaved landscapes.

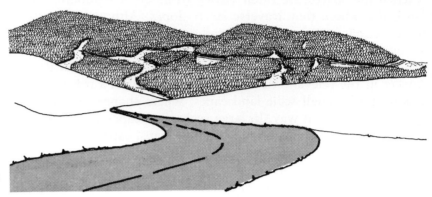

Fig. 13.19. Roadside spaces on a forested slope, included in an overall pattern of open spaces.

shape of this and other spaces can be designed initially from the most sensitive views and then adjusted from subsidiary viewpoints. Only where there are irreconcilable conflicts should either view be sacrificed, but the visual impact on the wider landscape usually takes precedence over the view from the road.

Roadside edge treatment

All the variables of detailed shape, spacing, pruning, and thinning described in Chapter 8 should be applied to the roadside edge at an appropriate scale. The appearance of the forest edge usually takes priority over the needs of wildlife in these locations. Spaced tree groups can emphasize the illusion of varying speed created by enclosed spaces close to the road.

Carefully positioned groups in the roadside, in combination with pruned and thinned edges, are simple and effective. Pruning should be done to varying heights, based either on tree diameter or on landform. Avoid an even height of pruning or a horizontal canopy level. Pruning should not exceed two-thirds of the tree height; trees from 4 to 10 m in height should not be pruned beyond one-third tree height, and smaller trees should be left unpruned. Treatment should follow landform, pruning to greater height in hollows or where the edge recedes from the road, reducing irregularly to nothing on summits where the forest edge is nearer. If the forest edges do not follow landform and cannot be adjusted to do so, a greater degree of difference in pruning height is needed between one side of the road and the other. On the lower side of the road trees will then be pruned to a maximum height near summits and not at all in hollows.

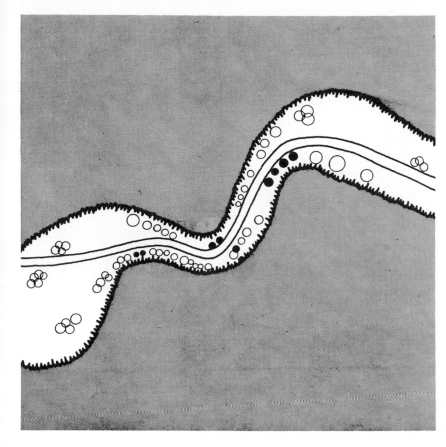

● Standard trees

⊘ Tree groups

Fig. 13.20. Tree groups positioned close together as the roadside space contracts near a sharp bend, giving an illusion of high speed.

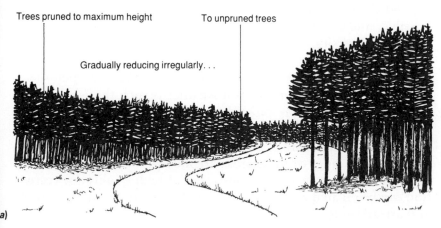

Trees pruned to maximum height

To unpruned trees

Gradually reducing irregularly. . .

Fig. 13.21. (a) Where roadside edges follow landform, trees can be pruned higher in hollows, reducing to nothing on spurs. (b) (see p. 294) Where evenly placed roadside edges are required the upper side of the road can be pruned higher in hollows and the lower side pruned higher on spurs, so that the pruned edge alternates from one side of the road to the other.

a)

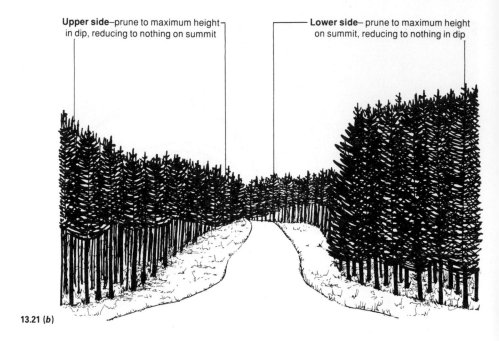

Upper side–prune to maximum height in dip, reducing to nothing on summit

Lower side– prune to maximum height on summit, reducing to nothing in dip

13.21 (*b*)

Recreational sites and walking routes

There are many areas within forest and woodland capable of development into very attractive recreational sites. They are a small proportion of the forest area, and because they are accessible by forest roads, now or in the future, they are valued by older people or families with young children who find long hill walks beyond their abilities. Such sites should be identified as early as possible, and incorporated into the landscape plan. They provide welcome diversity, even if they are not to be developed in the immediate future.

Good design can do wonders on recreational sites. Open space is very important because British weather is such that shelter and sunshine is needed more often than shade. The forest edge should be designed to enclose space at the appropriate scale while allowing some views out and providing a diverse, small scale landscape. If the surroundings of these areas are incorporated into an overall forest design plan, a mature landscape of high quality for recreation can be created at reasonable cost. These sites need easy access and attractive features, e.g. water, rocks, old trees, varied vegeta-

Fig. 13.22.

tion, and small scale landform. The appearance of woodland seen from car parks, picnic sites, etc., and their associated trails and walks should also be given careful consideration in the landscape design.

Archaeological sites may have a recreational role as well as being of historic interest, providing opportunities for interpretation and contributing to diversity. These sites should be safeguarded from disturbance by forest operations and open space associated with them should be included in the overall landscape design. It may be appropriate to reflect the historic context of the more interesting and important sites in the landscape design, in term of forest shape and species. The *genius loci* should be identified and carefully conserved. It is essential to seek professional advice when remains are suspected or found.

In England and Wales there is a range of designated rights of way over public footpaths, bridleways, and 'roads used as public paths', which landowners are obliged to keep open. In Scotland there is no statutory designation though some public rights of way have been established by court action. It is good practice to protect traditional walking routes, whether designated or not, especially between settlements or public roads and open hill land above the forest.

Walking routes are similar to motor roads in that the traveller sees a succession of small scale landscapes (see Fig. 13.22), albeit at much slower speed. Many of the principles of roadside land-scaping apply, with greater attention to small scale detail in the woodland edges. Statutory footpaths and bridleways follow a precisely defined route and the open space created by this line imposes a shape on the wider forest landscape. In some cases, it can appear intrusive, especially where it runs straight or at right angles to the contour. It may be necessary to plant trees close to the path at the most prominent points and to divert the associated open space along more sympathetic lines.

The methods outlined in Chapter 9 for the improvement of rides can be used with good effect to make walking and horse-riding routes, and cross-country ski trails more attractive. A simple seat or two at a major viewpoint is much appreciated (see Fig. 13.23), as are unobtrusive glades for rest and refreshment.

Fig. 13.23.

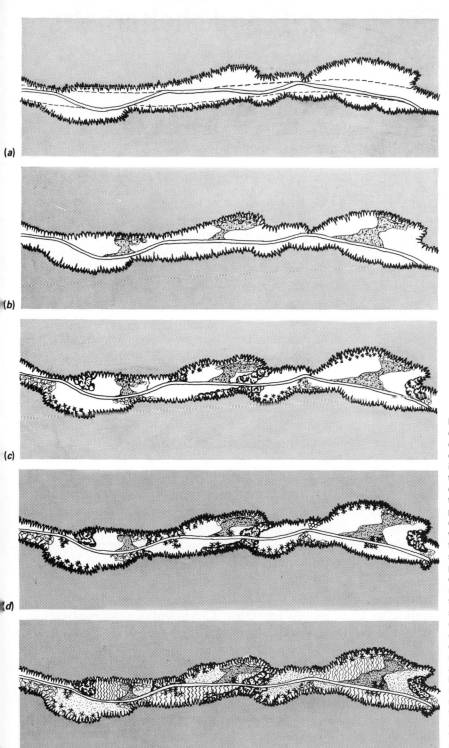

Fig. 13.24. Sequence of design for converting a ride to a glade for the enjoyment of visitors, as a picnic site or other informal use, and with some improvement of wildlife habitat. (*a*) Shape the crop edge to vary the size, width, and direction of spaces; vary the alignment of the path within the original ride. (*b*) Select places with good aspect or views for picnics or rest, and plant a thicket of large shrubs as wind-break shelter and to limit approach from the rear. Identify areas where grass should be cut to provide access to these points from the path. (*c*) Emphasize constrictions of space to varying degrees by retaining conifers and planting broadleaved trees and shrubs. Emphasize expansions of spaces by heavy thinning and pruning of edge trees where the glade is wider. (*d*) Add occasional tree groups where the path crosses extensive spaces, i.e. more than 40 m, and to discourage visitors from cutting across bends in the path. Identify edge broadleaves in groups, which should have high aesthetic quality and contrast with the surrounding mass of the woodland. Enhance forest stands close to the path by thinning and pruning. (*e*) Further diversity, both visually and for wildlife, can be achieved by managing other areas as low shrubs, coppice, or as rough grass.

(*a*)

(*b*)

(*c*)

(*d*)

(*e*)

14 Small woods and shelterbelts

- The design of small woodlands should reflect the degree to which landform or hedgerow pattern dominates the landscape.

- In upland landscapes small woods should be irregularly shaped to reflect landform and create an impression of large scale.

- Where the pattern of hedgerow trees dominates, a smaller scale and more geometric layout can often be adopted. If in doubt, follow landform.

- Irregular patterns should be created in the woodland edges in hedgerow dominated landscape.

- Even in the lowland hedgerow landscape avoid geometric shapes within the wood and especially close to skylines.

The general principles of landscape design can be applied to small woodlands, but there are additional complications of scale. It is wrong to imagine that the small size of these woods means that they have less visual impact and, therefore, need little design effort. In the larger upland landscape our attention is drawn by their very smallness. The dominance of farming often limits the amount of land available for trees and the resulting small scale of woodlands often conflicts with the large scale of the landscape. The location of new woodland in the agricultural landscape is strongly influenced by land quality, ownership pattern, methods of husbandry, and previous investments in agricultural improvements, such as re-seeding of pastures, but it is, nevertheless, most important that the needs of landscape and wildlife are taken into account in small woods on farms. Landowners must be aware of the nature of these requirements before any work is planned or begun.

Landscape appraisal

The key aspect of landscape appraisal for small woodlands is whether landform or the pattern of trees and woodlands dominates

Fig. 14.1. Landform readily becomes the dominant visual influence where hedgerows have been extensively removed.

visually. Small woodlands are an essential element in the agricultural landscapes traditionally associated with lowland Britain. It is, therefore, relatively easy to introduce additional woodlands to the small scale framework of hedgerow trees on flatter ground. If a rectangular field pattern is translated as woodland to the larger scale, steeper, and higher slopes of the uplands, the whole geometric shape of the wood becomes visible and is seen to be quite inappropriate.

Even in lowland and downland landscapes, landform becomes visually dominant where hedgerows have been extensively removed. Here woodlands should be planted on a larger scale and shaped to reflect landform, rather than attempt to recreate piecemeal the enclosures of previous centuries. The latter would appear out of scale unless comprehensively developed over large areas. Larger, well-designed areas of woodland are also easier to manage, for a variety of objectives, in conjunction with the farming enterprises on adjoining land.

Small woodlands in the hedgerow landscape

The traditional agricultural landscapes of lowland England and Wales of fields, hedgerows, trees, and small woods has widely recognized aesthetic qualities. If this landscape is to continue, it must be conserved and renewed. In this context the shape of small woodlands need not follow landform closely where the hedgerow pattern dominates. New woods should, however, reflect the scale and irregularity of the pattern, while interlocking strongly with open space. Achieving this is not easy because the visual texture of plantations, even broadleaves at wide spacing, is much finer than many hedgerow landscapes. On flat ground the even height of canopy creates a horizontal line, quite different from the irregular crowns of spaced hedgerow trees. On lowland hillsides young plantations still appear as shapes sufficiently distinct from hedgerows to create a geometric appearance. Irregular shapes and edges are, therefore, needed to reflect the coarser texture of surrounding hedgerows.

Fig. 14.2. Diverse fields, interlaced with an irregular pattern of hedgerow trees and small woodlands, on gently rolling landform in the Malverns, Worcestershire.

Fig. 14.3. A more geometric and introspective pattern of continuous even-aged shelter woods in Peebles-shire. Such a pattern would be inappropriate in the more irregular southern English landscape.

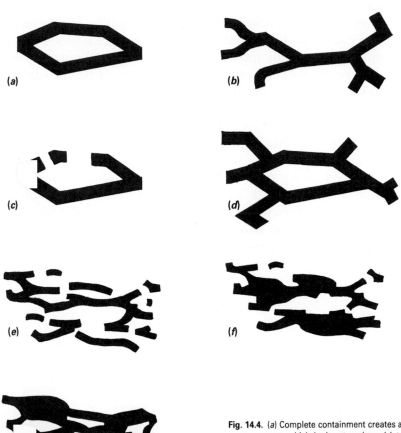

Fig. 14.4. (a) Complete containment creates an inward looking space. (b) A branching shape creates spaces which look outwards and interlock well with the mass of woodland. (c) An introspective enclosure may be unified with the landscape by partial opening or (d) by extending into the surroundings to create a more outward-looking landscape. (e) A careful balance of enclosure and openness can combine and blend a small scale space with more open surroundings. (f) Additional planting can either increase the interlock or (g) reduce it.

Enclosure, scale, and unity

The incompleteness of hedgerow enclosures creates a strong interlock of woodland and open space which contributes to the character and quality of many agricultural landscapes. It gives a highly unified appearance in the distance; a succession of views are seen from within, framed by trees to form reassuring enclosures on a human scale. The more outward-looking character of these partial enclosures blends them more strongly with their surroundings than completely enclosed fields as seen in the Scottish Borders. As a result the more open appearance is more easily unified with open ground at higher elevations or where the tree pattern breaks down.

These aspects of enclosure can be used to guide the development of an existing woodland pattern. Where the agricultural landscape is more open or adjoins open hill, a generally 'branched' shape with outward-looking spaces will blend better with the surroundings. Existing partial enclosures should not be filled in by planting new woods which create inward-looking spaces separated from the wider landscape. By positioning woods in an outward-branching pattern it is often possible to plant a large total area while maintaining an intimate scale of landscape.

An interlocking pattern is also more effective than an even scatter of clumps which appear too small in scale, especially near the skyline. Scattered woodlands often appear to be awkwardly separated from the landscape, 'adrift', and limited additional planting should, if possible, link such areas into a pattern of broader scale.

Fig. 14.5. (a) An area of low-grade broadleaves appears disconnected from the general hedgerow pattern. Powys. (b) Small areas of additional woodland would strengthen the overall woodland pattern, improving scale and reflecting landform on the upper slope.

(a)

(b)

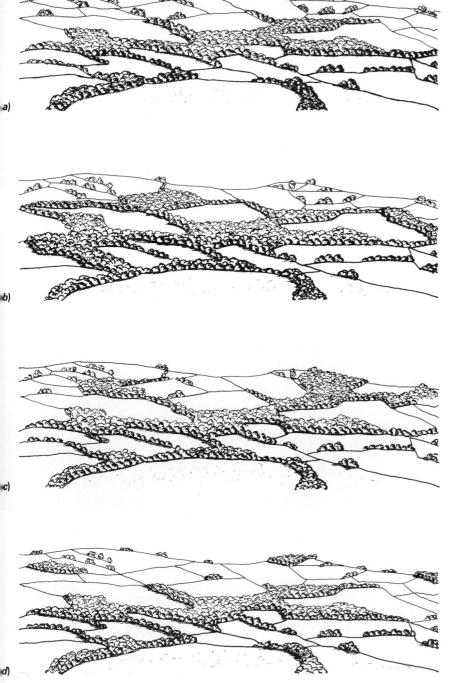

Fig. 14.6. Extending farm woodlands in a hedgerow landscape (1). (*a*) 'Branched' pattern of woodland well unified with the hedgerow landscape. (*b*) Additional planting which closes off spaces, separates the woodland pattern from its surroundings, and is too large in scale. (*c*) Planting which extends the interlock of open space and woodland maintains the intimate scale of the pattern and the unity of the landscape. (*d*) Scattered woodlands may be too isolated to contribute to the overall pattern, and may appear untidy.

Where trees in the hedgerows are more dominant, the location of new woods is less critical, though planted fields should not be amalgamated into large geometric shapes so as to intrude on the scale of the broad pattern. Here again, additional woodland can be more sympathetically introduced if interlocking shapes are

(a)

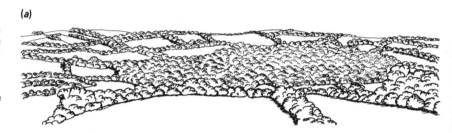

Fig. 14.7. Extending farm woodlands in a hedgerow landscape (2). (a) Although the choice of areas for further planting is less critical where the hedgerow pattern is stronger, large-scale geometric shapes should be avoided. (b) (opposite) A more interlocking shape maintains the intimate scale of the landscape while allowing a larger area to be planted.

Fig. 14.8. (a) Enclosure pattern and strong landform finely balanced; the long horizontal margin of the central wood appears too straight at this scale. Additional planting to improve this boundary could follow either the landform or the hedgerow pattern. Forest of Dean. (b) Additional planting following landform. (c) (opposite) Additional planting following hedgerow pattern.

(a)

(b)

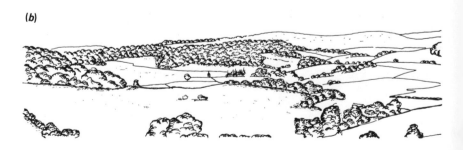

adopted. Any hedgerows completely within the new woodland should be partially felled, if necessary, to create the irregular effect more appropriate to the interior of a wood, and any resulting coppice regrowth should be incorporated into the species pattern of the wood.

(b)

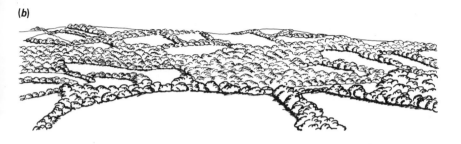

(c)

Shelterbelts and small woods in the upland landscape

Small woods for sporting, shelter, and timber production which readily fit into a lowland landscape with a strong pattern of hedgerows often appear too small in upland and more open landscapes. Small woods should only be planted where the small scale is appropriate. This may be close to existing woodland or hedgerows; on lower slopes of valleys; or near (but not screening) other landscape features such as buildings, crags, watercourses, or other features.

Shape and the contrast of woodland with open hill are every bit as important in the design of small woodlands as they are for

Fig. 14.9. Rectangular, symmetrical design is no less unwelcome because the areas are small. Although the shapes and spacing of these woods could easily be improved, their intrusive small scale in this large landscape remains a problem.

Fig. 14.10. Small isolated woodlands appear out of scale in such open landscapes. As the number of woods increases, the effect becomes worse.

extensive forests. The geometric shapes advocated for shelter and game cover in the past have to be substantially modified to blend sympathetically with the landscape.

The shape of small woods and belts should vary in relation to landform, rising in hollows and falling on convex ground as described in Chapters 6 and 7. The width of woodland belts should be irregularly varied, and they should have enough breaks to interlock visually with the open space as outlined in Chapter 8.

As the number of separate woods increases, the small scale of their overall pattern is emphasized, often in conflict with the larger

Fig. 14.11. Planting partially encloses an open space, giving an impression of a much larger woodland area than is actually planted. The shape is much too geometric.

Fig. 14.12. An impression of continuity can be achieved if the woods are close enough together. Even a single tree can affect this impression in the smaller upland landscapes.

landscape. Design is further complicated by the greater interaction of different shapes. An impression of extensive interaction of different shapes. An impression of extensive woodland can be created by either planting woods, so that they partially enclose larger spaces or by locating woods so that they seem part of an overall pattern.

Small woods which seem to overlap or are sufficiently close together give an impression of greater scale and continuity than when they are scattered. The apparent space between neighbouring woods is critical in maintaining the impression of continuity. The presence of even a single tree can be significant.

The devices of closure, overlap, and nearness can be used in various combinations to give an impression of larger scale. Overlap and nearness are generally more effective on gentler convex slopes, and in foreshortened views, while closure is more appropriate on steep slopes and where there are important views from the space between woodlands.

Prominent skylines should not be outlined by isolated belts unless supporting woodland extends down the hill face in places, to improve scale. Belts ending abruptly on the skyline, or planted just below with a sliver of open ground above, appear out of scale and should be avoided. Woods should either encompass the skyline in a substantial way or be kept well clear of it. Deteriorating shelterbelts on the skyline look awful; if they cannot be maintained they are better felled on a large scale and replanted to a good design.

Fig. 14.13. Woods and farmland where the interlocking pattern of diagonally curved shapes is a good example of well-designed small upland woods. Broughton, Peeblesshire.

(a)

(b)

Fig. 14.14. (a) As well as being too regular, this belt framing the skyline has an intrusive scale which could be improved by closure and overlap. (b) The same belt would look better with additional enclosing and overlapping woodland.

(a)

(b)

(c)

Fig. 14.15. (*a*) Shelterbelts should not end abruptly on prominent skylines but (*b*) should curve gently round them. (*c*) Shelterbelts should not frame the skyline without (*d*) supporting woodland on the front slope, creating overlap and enclosure. (*e*) If shelterbelts are not planted on the skyline, they should be kept well clear. (*f*) Shelterbelts close to the skyline give intrusive slivers of open ground and the wrong scale.

(d)

(e)

(f)

Small woods should not be placed symmetrically on hills; this looks very artificial. The impression of symmetry varies considerably from one view to another, so all public viewpoints should be carefully identified in advance.

Fig. 14.16. (*a*) A symmetrically placed belt which looks both artificial and intrusive from the public road. (*b*) When positioned away from the summit of the hill, the wood immediately appears to be a more incidental part of the landscape. (*c*) Greater variation in width improves the appearance further.

(*a*)

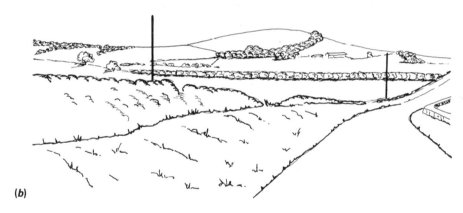

(*b*)

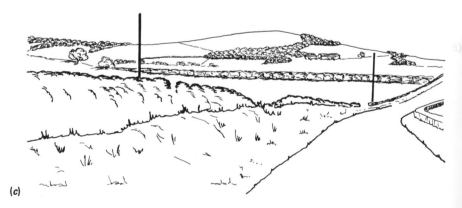

(*c*)

Occasional small clumps of woodland can be planted to emphasize points of interest in the landscape or landform. Prominent positions should be planted sparingly, otherwise the eye becomes confused by small points competing for attention. Where additional clumps are needed for game cover, they should be varied in size and spaced irregularly apart. They will also appear less intrusive if viewed against the background of a larger wood.

All these points apply wherever lowland landscapes are dominated by landform rather than hedgerows. The device of overlap can be used to great effect on gentler slopes to enable the woodlands to reflect the often unexpectedly large scale of the open lowland landscape.

(a)

(b)

Fig. 14.17. (a) Woodland belts positioned rather too far apart to create an impression of continuity in scale with the landscape of the Lambourn Downs, Berkshire (courtesy Pictor International). (b) With additional planting the scale of woodlands would reflect the landscape better.

Hedgerows and landform in the uplands

Hedgerows are found on the lower slopes of many upland valleys, and, where the pattern is strong, more regular woodland shapes can be used on gentler slopes. Do not underestimate the influence of landform: wherever the pattern of hedgerow trees weakens and gives way to walls, fences, or deteriorating hedges, landform becomes visually dominant. If in doubt, shapes should follow landform, especially closer to open hill and skylines. Woodland should be planted as an interlocking element between the open hill and the hedgerows below.

Edge treatment of small woodlands

The visual factors affecting edges of small woods are the same as for larger forests, though the need to develop diversity is less acute. Edge design may have to be modified to meet the requirements of shelter or game. In the lowlands the smaller scale of the hedgerow landscape means that detail is important. The irregular distribution of trees along the hedgerows should be reflected and emphasized in the woodland edge.

Existing hedgerow trees in the woodland edge are of great value to landscape and wildlife. They should not be prematurely felled and their crowns need space to develop. Planting should be kept far enough away so that young trees do not compete for light, or overshadow and so hide the crowns.

The size of large individual hedgerow trees is such that the loss of a few individuals has a major impact on the local landscape, as was illustrated by the effects of Dutch Elm Disease. The woodland edge should, therefore, be planned so that younger groups of trees growing in the open can succeed the present mature giants in due course. The even age and height of a plantation makes it difficult to match the irregularity of a mature hedgerow, so the diversity of the woodland edge should be exaggerated. The amount of irregularity in the hedgerow pattern in the vicinity of the wood, the general density of trees in the hedgerows and the variation in their spacing indicate how woodland edge groups might be distributed.

The continuity of the hedgerow pattern is easily lost when adjoining woodland is felled, unless some individual trees and groups are retained. Whippy, one-sided trees are unsightly, so it is best to delect trees to be retained well in advance of felling and give them space so that their crowns develop fully.

Some ancient hedgerows are of great historic and botanical interest, often characterized by a wide range of tree and shrub species. It is important to recognize these survivals and to ensure that their special features are conserved.

Woodlands for game

A balanced approach to planning and management is necessary to achieve a satisfactory appearance for woods planted for game. The principles to follow in their design are those described above, as

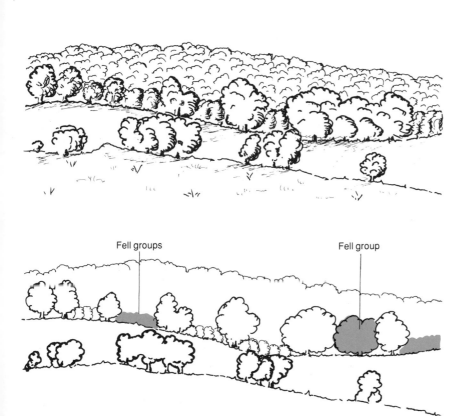

Fig. 14.18. (a) Where the density of hedgerow trees beside the wood is typical of the surrounding landscape, some outlying groups should be planted in between hedgerow trees, of which a few should be replaced by planted groups to perpetuate the pattern. (b) (see p. 316) Where there are fewer trees beside the wood, indentations in the edge and outlying groups should be used to enhance diversity. (c) (see p. 317) Where tree density is higher, selected groups of hedgerow trees should be felled and replanted in a pattern more typical of the surrounding hedgerows.

(a)

the effect on the landscape is the same, whatever the reasons for planting.

On steeper ground and with larger areas of woodland, the shape of the layout is increasingly important. The shape should reflect the form of the ground so that edges and boundaries rise uphill in hollows, and fall on convex slopes. Boundaries should not run straight up slope or follow contours horizontally, and right angles should be avoided except on flatter ground where traditional enclosure boundaries dominate.

Asymmetric curved diagonal shapes with varied width of belts represent the ideal. Optimizing management tends to favour

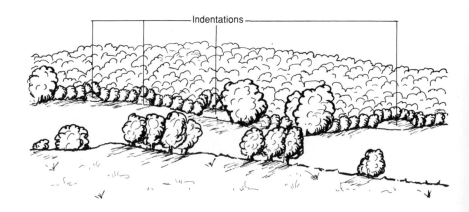

14.18 (*b*)

straight-fenced compact shapes, but with a little imagination the formality of the square and rectangle can be avoided. Trapezoids, scalene triangles, and other shapes with sides of varying length offer a possible compromise of informality and economy.

The game covert edge usually contains large conifers for shelter, medium sized broadleaved trees and shrubs for food, and a hedge of some sort to keep out the wind. Planting these constituents in continuous belts gives a very formal series of parallel layers, so (with the exception of the hedge) they should be laid out as irregular overlapping groups, as described in Chapter 8.

Groups for felling

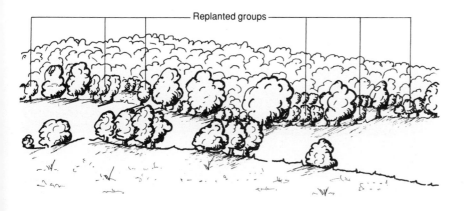

Replanted groups

14.18 (c)

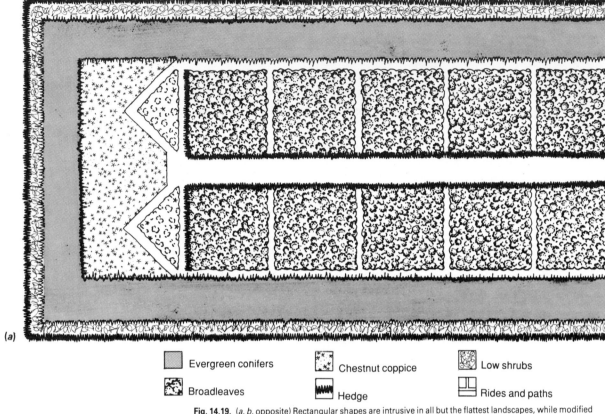

(a)

Evergreen conifers   Chestnut coppice   Low shrubs

Broadleaves   Hedge   Rides and paths

Fig. 14.19. (*a, b,* opposite) Rectangular shapes are intrusive in all but the flattest landscapes, while modified layouts lie more easily on the contours.

Fig. 14.20. Layers of different species in the edge of a pheasant covert form very artificial continuous belts of trees.

14.19 (b)

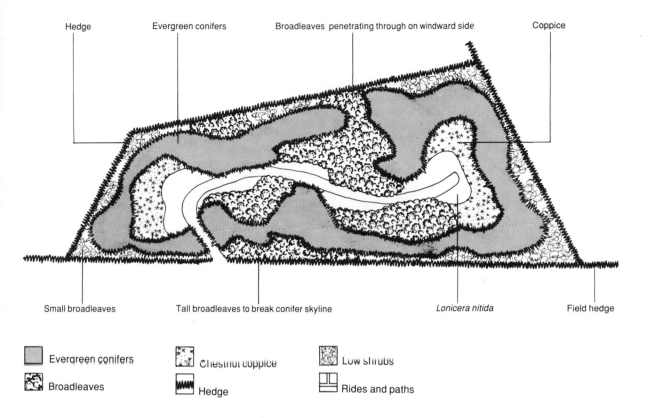

Hedge   Evergreen conifers   Broadleaves penetrating through on windward side   Coppice

Small broadleaves   Tall broadleaves to break conifer skyline   *Lonicera nitida*   Field hedge

Evergreen conifers

Broadleaves

Chestnut coppice

Hedge

Low shrubs

Rides and paths

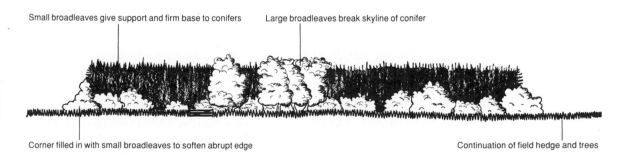

Small broadleaves give support and firm base to conifers   Large broadleaves break skyline of conifer

Corner filled in with small broadleaves to soften abrupt edge   Continuation of field hedge and trees

Fig. 14.21. An asymmetric pattern of overlapping groups results in a more natural appearance.

Landscape design process

- The broad and detailed influences of the landscape on the forest area are carefully appraised.

- Existing and potential aesthetic problems should be identified and solved.

- The design concept for the forest should be derived from the broad landscape influences, operational, and environmental requirements.

- As many natural details as possible should be incorporated into the design.

- The possibility of refinement and adjustment should be maintained for as long as possible during the design process.

- The visual quality of the design should be checked by accurate illustrations from all relevant viewpoints.

This chapter provides a guide through the process of design so as to ensure that comprehensive information is gathered and organized in a useful way.

The designer must obviously bear in mind the goals of design and the pursuit of excellence;

(1) to reflect and enhance the best natural qualities of the landscape while providing occasional contrasts;

(2) to eliminate visual intrusion and achieve outstanding design in the sensitive parts;

(3) to develop a range of diverse habitats;

(4) to balance all the above factors with operational efficiency.

The design process falls into three main stages; information gathering (landscape survey), information sorting (analysis), and design synthesis. It is difficult to know in advance all the information which will be useful when it comes to the design synthesis, so all remotely relevant information which is readily available should be recorded. Throughout the process it is important to look first at the large-scale, and most visible influences and effects, and to fit the details into the context later. If the broad pattern is forgotten, the resulting design will be cosmetic pastiche.

Landscape survey and analysis are combined into **landscape appraisal** which is the basis for developing a design concept.

Landscape survey

The checklist below is divided into: broad influences; elements of diversity; aesthetic problems; and management and other site factors. The broad influences are central to the development of a design concept, during the formulation of which solutions to aesthetic problems, factors of management, and elements of diversity are incorporated as far as possible.

To ensure that broad influences are properly considered, resist the natural desire to travel direct to the site. Many of the broad assessments should be made in a regional context and should be applied to the whole landscape within sight of the forest area. First plan different approaches by a variety of means, using a 1:50 000 scale map. The most distant of the possible views of the site are visited first and photographed. As well as the views which will be seen by the majority of people, it is useful to record others from surrounding skylines and points from which the design can be readily comprehended. They provide a good impression of the general location, landform, and space of the site. Roads through the area should be travelled in both directions, from where the area first appears to where it is last seen. This should initially be without stopping to assess the influence of the area on roadside views; specific points can be recorded later.

Increasing amounts of detail are recorded as the landscape survey proceeds. The depth of research will be influenced by the importance of specific features and how easily they can be included in the design. More detail is generally required where there is potential for conflict.

Photographs record a large amount of information quickly and provide the most objective base available for sketch design. Although photographs distort the image of the landscape slightly, they make it much easier to assess the aesthetic influences without being distracted by detail. Cost of film and processing should not inhibit comprehensive recording; it will be less than the cost of a return visit to check unrecorded details.

TABLE 15.1 Landscape appraisal checklist

(A) Broad landscape influences

Consider the importance, value, and effect of forest operations on the following factors, and how they might influence design.

Factor	Information required or to be considered	Chapter	Methods, sources, and references
1. Landscape sensitivity			
status of area	National Park AONB Area of Outstanding National Beauty NSA National Scenic Area AGLV Area of Good Landscape Value other local classification	4	Structure plans Scotland's Scenic Heritage (1978) local plans
quality of	Degrees of diversity Degrees of unity Presence of *genius loci* Presence of water Presence of eyesores	4	Site visits Photography
visibility of area	General level of population Local residences Recreation use of areas in sight Road status and traffic flows Slope Elevation Views inward	4	Library, local authority Site Site, map and local authority Map and local authority Contour map and site Contour map and site Map and site
2. Heritage value	Archaeology Associations with historic events Associations with literature Association with works of art Designed landscape Traditional events or religious Significance Traditional patterns of land use Extent and rarity of the above areas	4, 5 4	Local library Local historical society Local art gallery Site, photography Site, photography
3. Feature and background	Regional context: is site typical or atypical Local context: is site typical or atypical	4	Site visit, photography Landscape descriptions, e.g. in plans or Scotland's Scenic Heritage

Dominant features:
 peak
 pinnacle
 waterfall
 cliff
 entrance
 skyline
 building
 settlement

'Background' areas:
 extent
 character

Main views:
 quality
 visibility
 sequence

4. Landscape character		
Natural Components		
Which of the following combine to give the landscape its essential character?		
Landform scale shape direction texture	4	Site visit — photography, aerial stereo pairs of photographs
Vegetation texture pattern colour	4	
Man made attributes		
Land use pattern clearings shelterbelts treed hedgerows hedgerows without trees walls fences woods	4	Site visit and photography — aerial photographs
Settlement pattern (existing and proposed) scale distribution form		Site visit, maps, structure, and local plans
Vernacular buildings and structures grouping spaces	4	Site visit and photography

Factor	Information required or to be considered	Chapter	Methods, sources, and references
Man made attributes (cont.)	form materials texture pattern (windows) colour details		
Aesthetic Factors	Describe local scale shapes diversity unity rhythms texture	4	Site visit and photography

5. Spaces	Whether defined by woodland or landform
	  exposed
	  open
	  enclosed
	  confined
	  oppressive
	  canopied
	  expanding
	  contracting
	
	Sequence of enclosure and openness, especially along roads and paths; proportion and balance of woodland and open ground; degree of definition — diffuse or crisp
			Contour maps; site visit — sketching and photography; Erno and Goldfinger (1940). The sensation of space. Architectural Review, 90, 129–31.

6. Sequence	Routes and destinations Arrival Gateway Climax Anticlimax Diversity Sense of motion Unified or disjointed Varied or repetitive	4, 13	Site observation, photography, sketches Appleyard et al. (1984) Driving and/or walking route

Factor	Information required or to be considered	Chapter	Methods, sources, and references
7. Aspect and climate	Identity dark dull cold areas Consider colour shadow shelter Winds — prevailing Frost hollows — cold	9, 10	Contour maps, compass
8. Spirit of place	Combinations of character identity structure features atmosphere unity diversity space things	2, 4, 5	Site and photography sketches

(B) Elements of diversity

Elements of diversity to be recorded on map and transferred to analysis photosketch if they are to be incorporated into the design.

Factor	Information required or to be considered	Chapter	Methods, sources, and references
1. Open space	Qualities textures colours Location farmland unplantable land important wildlife habitat beside roads, paths transmission lines	5 5 5, 9 5, 9, 13 5, 9	Site, photography, maps, aerial photographs

Factor	Information required or to be considered	Chapter	Methods, sources, and references
1. Open space (cont.)	Management requirements deer glades felling coupes nursery land car and vehicle parks timber storage	5, 9 5, 9, 12	
2. Views	Location from settlements, routes, or recreation areas Direction inward outward Type panoramic enclosed focal feature static/dynamic	5, 13	Maps, site, photography
3. Water	Still — reflections Moving Edge quality Habitat Catchment areas, water supply reservoirs, points of abstraction, fishing	5, 9 5, 9	Maps, aerial photographs, site, and photography Angling clubs
4. Landform	Scientific interest Character Smoothness Form	5 4, 5 4, 5 2, 5, 6	*Nature Conservation Review*, NCC (1977), site, photography, contour maps
5. Rocks, crags, and scree	Scientific value feature or characteristic; isolated or in patterns; area needed to avoid screening; Sites of Special Scientific Interest (SSSI), Nature Reserves.	4, 5 5 5, 9	*Nature Conservation Review*; site, aerial photographs, maps, photography
6. Vegetation	Scientific or rarity value of species and communities; large and old trees to be kept unscreened; ancient semi-natural woodlands; common plants of aesthetic value individually or in mass potential for interpretation; SSSI, nature reserves.	5	Site, *Nature Conservation Review*; NCC: Register of Ancient Woodlands; local knowledge, local natural history societies

7. Animals	Habitats for conservation or improvement; management areas viewing opportunities potential for interpretation; SSSI, Nature Reserves.	5, 9	*Nature Conservation Review*; local knowledge; local natural history societies; local keepers and rangers surveys and studies
8. Archaeological sites	Significance Date Extent Distribution Views inwards Potential for interpretation	4, 5	Government archeological departments; local libraries; local authorities (county archaeologist or equivalent); site, aerial photographs, maps
9. Recreation	Existing recreation areas and use	5	Site, photography
	Potential recreation areas in high quality small-scale landscapes.	5, 9	*The public in your woods* (1985).
	Aspect Demand Access		Contour maps, site; local and structure plans, maps, deeds, site
	Extent of recreation use Rights of way and traditional walking routes	5, 9, 13	Sites questionnaire Maps, sites, local plans ROW definitive maps
	Long distance paths	4, 5	Maps, site
10. Forest stands	Variations in age — exceptional areas and individuals Variations in species — exceptional areas and individuals Variations in management regimes: coppice coppice with standards ancient woodlands ancient semi-natural woodlands	5, 12	Stock maps, aerial photographs, site visit, and photography Register of Ancient Woodlands

(C) Aesthetic influences and problems

Where these problems arise they should be recorded on an appraisal map and photosketches before design begins.

Design principle	Likely landscape influence	Forest element	Chapter	Most common problems	Remarks
Shape	Landform Rocks Ground vegetation	External margins Legal boundaries Species margins Service lines Wayleaves Open spaces Felling coupes Rides	2 7 6	Straight lines, right angles, lines following contours; lines cutting across in contours; parallel lines of fine belts, fringe, retentions, open spaces	Describe and photograph, landscape influence; record problems on map and photosketch
Scale	Landform Breadth of view Distance of view Elevation Skyline Woodland and hedgerows on flatter ground	Forest block	2 7	Extensive uniform forest on small-scale landform; large uniform areas seen in short views	Scale is large to and from hill tops; scale small in narrower valleys especially on lower slopes Record problems on appraisal photosketch
				Small blocks high on hills in longer views	
		Upper margin	7	Too close to skyline leaving narrow strip of open land Appears over skyline as a narrow fringe of trees	
		Edges	8	Appear too uniform in near views	Consider views from roads, paths, and recreation areas and treat as a refinement of the design of margins
		Species margins	10	Looks too 'fussy' as a result of following ground vegetation patterns too closely	
		Felling coupes and other open spaces	12 14 8	Appearing too large in valleys; appearing too small near skylines; small numerous spaces appear 'moth eaten'	

Category	Description	Factor	No.	Problem	Action
Visual forces	Convex and concave landform and slopes, skylines	Broadleaved trees	13	Moth eaten appearance if scattered too widely	Draw hierarchy of upward and downward forces on contour map and transfer to appraisal photosketch (accuracy essential)
		External and species margins open spaces and rides	2	Straight lines cutting across contours; corners pointing across and against visual forces in landform; margins rising on spurs and falling in hollows	
Diversity	Patterns of open space, natural features; general level of diversity in surroundings	Canopy, species age	2, 5 10 12	Natural features hidden by tree canopy Large areas of evergreen species appear out of place and uninteresting in more diverse surroundings Large areas of forest where deciduous degrees cannot be established: lack of open space	Record factors at B, above, to be left clear of planting; identify areas which might grow deciduous species Identify additional areas to be left unplanted; illustrate complete cover on photosketch to assess appearance
Unity	Local shapes and general qualities assessed as above; pattern of woodland an open space	Complete forest	2 9	Continuous open spaces, e.g. rides and roads breaking up forest excessively Forest does not blend with or reflect shapes and scale of surroundings Abrupt edges	Use interlocking shapes of forest and open space, and eliminate rides

(D) Management and site factors

The relevant and important factors should be recorded on map and main photosketch before design begins. Most of it should be obtained in the briefing meeting.

Factor	Relevance to Design	Chapter	Sources and references
Economic limit of planting	Indicates approximate position of upper margin	7	Exposure and soil maps
Windthrow hazard	Location of felling coupe boundaries	12	Exposure and soil map
Limits of utilizable timber	Location of contrasting species	10	Exposure, elevation and soil map
Location of utilizable timber produced by broadleaved species	Location of contrasting species	10	Exposure, elevation and soil map
Areas to be thinned	Accessibility for harvasting shaped coupes, thinned coupe edges, improved edges seen in shorter views	11	Compartment records, yield class, windthrow hazard class stock map
Economic felling dates	Planning of phased felling	12	Compartment records, stock map
Dates at which critical height is reached		12	
Rate of greening over' of restocking	Timing of phased felling	12	Local knowledge and observation; soils map
Planned volume production	Rate of felling	12	
Methods of extraction	Accessibility of different areas of shaped coupe	12	Terrain classification local knowledge
Harvesting roads	Accessibility of different areas of shaped coupe	12	Roads map/stock map
Dear management	Distribution of open space	9, 12	
Ownership boundaries	Limits of external margins or identifying where additional acquisition is required	7, 12	Title deeds
Agricultural land	Limits of new planting	7, 9	
Recreation areas, forest walks, rights of way	Scale and impact of felling, re-routeing while felling is in progress; safety	9, 12	Local knowledge, local authority definitive maps
Areas used for rearing of game birds	Scale of felling, replanting species	12, 14	Local knowledge

Analysis of landscape survey

The object of analysis is to organize the recorded information so that it can be used readily in design synthesis. Information can be classified as:

objectives	constraints
influences	problems or weaknesses
assets or strengths	threats
opportunities	conflicts

It is important to distinguish between external factors such as constraints which cannot be changed and are to be avoided if possible or accepted; and internal factors which are under the control of management and where a solution is possible.

Information can be ranked according to the extent of its effect on the landscape and its importance. The two are not necessarily the same; factors identified in the survey should be ranked, first, in order of importance and, then, in order of their extent. Their importance indicates how much effort should be put into solving any associated problems, and extent indicates how soon attention should be given to them during design.

Take care not to undervalue the broader influences of the landscape, which may be more subtle than immediate visual problems. Solving the latter is often thought to be more important than taking advantage of a good opportunity, but the designer cannot afford to miss an opportunity, if any areas of excellence are to be achieved.

Different factors may interact because they affect the same place or because they can be co-ordinated. The same open space can be used for watercourse protection and for deer management. Factors which interact on site can be identified by the use of transparent overlays or 'sieve maps' positioned accurately on a base map. They allow details of different in formation to be compared without the overall appearance becoming too confusing. Alternatively, factors affecting the design can be set out in a simple matrix and interactions identified.

Combinations of different factors often create the unique quality or spirit of a particular landscape. With so many aspects of the landscape to consider, the most important must be clearly distinguished if a manageable design is to be established. Once the range of factors has been identified, the ranking described above should be carried out ruthlessly to determine priorities.

Design synthesis is the most difficult part of design, to carry out and explain. It is essential to recognize the large issues and problems, and deal with them at an appropriate scale, then tackle the less prominent and smaller-scale issues later. The designer must be open-minded about changing various parts of the design for as long as possible. The process of design may be seen as a cycle of creation, expression, testing, revision, and adjustment, with each design drawing critically assessed and refined. While all aesthetic aspects have to be taken into account, the design of forest shapes is consistently found to have the greatest influence on the quality of design.

The first step is to establish a valid design concept, derived from the broad landscape influences of the area. If a concept is not readily apparent after careful study of the landscape appraisal, ideas can be stimulated by:

(1) contemplating the landscape;
(2) studying photographs, aerial photographs, and maps;
(3) making rough sketches of the existing landscape;
(4) discussion with colleagues;
(5) looking at other design solutions;
(6) leaving the problem alone and coming back to it later.

Photosketches of the main views are the basic design documents, together with maps and aerial photographs. The concept is drawn in appropriate shapes, following the principles of design and maintaining a sufficient level of open space, especially near water. Aesthetic problems should be solved at this stage. It is essential to use accurate perspectives to display the design, to illustrate the problems to be solved, and features to be kept clear to trees. The techniques for doing this are described in the next chapter. The order of design should be as follows.

Afforestation of bare land

1. General location of woodland and open space
2. External margins   illustrate in summer foliage
3. Open space within the forest
4. Location of broadleaved trees

5. Species pattern and margins   illustrate in autumn foliage

6. Detailed edge treatment   illustrate shorter views

Design of existing forest

1. General location of woodland and open space ⎫
2. Improvements to external margins* ⎭ illustrate in summer foliage
3. Shapes of felling coupes
4. Timing of felling coupes
5. Visual effects of timing ⎫
6. Restocking species ⎬ one drawing for each period of felling, showing increasing growth of trees in each area
7. Detailed edge treatment — illustrate shorter views

* If few adjustments are needed, this may be done after 5.

It is important to test the design from different viewpoints. If possible, the design should look right from every viewpoint, but where there are unresolvable conflicts the most sensitive viewpoint should be given priority. Adjustment for subsidiary views is rarely a cumbersome task if the shapes are designed at the right scale and are well related to landform.

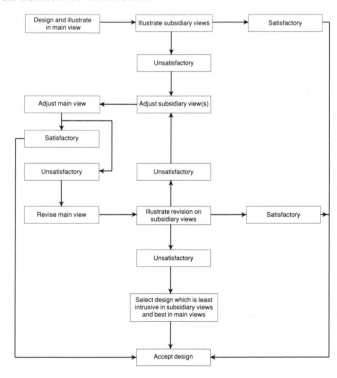

Fig. 15.1. The process of adjusting a design to accommodate different viewpoints.

16 Design techniques

Landscape plan documents

A landscape plan requires drawings, maps, and photographs for:

(1) recording landscape survey details;

(2) illustration of design from main viewpoints;

(3) accurate recording of boundaries and measurement of areas;

(4) accurate setting out of forest shapes on the ground;

(5) monitoring and adjustment on implementation.

Panoramic photographs provide an objective record of the landscape and are sufficiently life-like to be easily identified by members of the public.

Perspective sketches are essential for testing the visual aspects of the design, comparing options, presenting proposals, and for publicity. They are best based on panoramic photographs, which are more accurate and easier to produce than free-hand sketches.

The best method is to photocopy photographic prints of the landscape. The photocopier must be capable of accurately reproducing a wider range of half tones than the normal office copier designed for high contrast. Laser-based copiers are now available, with the ability to reproduce a wide range of half tones from colour or monochrome photographs, and are worth considering if the volume of work is large.

Alternatively, a sketch can be produced from a 35-mm transparency by tracing from an image projected onto a sheet of paper at a distance of about 1.5 m. If the image is larger than A3 size, use tracing paper; copies can then be produced by dyeline printing more cheaply than photocopying. Main ridge lines, valleys, and other dominant features are sketched first using a sharp pencil or fine felt-tip pen. Patterns of ground vegetation, etc., can then be blocked in with a broader felt-tip or soft pencil. Use red or brown colours where dyeline printing is intended. The drawing produced can then be copied and used for design, with the original image projected onto it as necessary to provide additional detail.

A third option is to draw on clear acetate sheets overlaid on

photographic prints. Fine-tipped spirit-based markers are necessary to avoid smudging. This is a quick method of dealing with
simple designs which can be readily agreed, but does not lend itself
to reproducing large numbers of drawings and details of the landscape background.

TABLE 16.1 Comparison of perspective bases for landscape design

Technique	Advantages	Disadvantages
Photocopying direct from prints	Reproduces accurate detailed base for design quickly and in large numbers, if necessary. Immediate graphic quality High credibility	Requires photocopier which will reproduce a wide range of halftones. This may cost more initially.
Tracing from a transparency	Accurate Readily available at low cost. Large numbers easy to reproduce by conventional photocopier or dyeline printer.	Detail lacking or time-consuming to apply unless slide is reprojected on sketch. Difficult to make composites because of need to reproject. Lacks immediate identifiable graphic quality unless slide is reprojected onto sketch.
Overlay on photograph	A lot of detail is visible if clear acetate is used. High degree of accuracy possible.	Detail lost when overlay is removed from photograph. Lacks graphic quality Unable to reproduce numerous copies. Clear acetate sheets expensive; cheaper substitutes hide detail and lose accuracy.

Photography for landscape design

The object of this type of photography is to show as much detail of the landscape as possible, especially topography, rather than to create artistic compositions or record interesting lighting.

The most appropriate camera is a 35-mm single lens reflex with through-the-lens metering. Models with manual setting of exposure and shutter are better than automatic cameras since landscape needs accurate exposure which has not been distorted by light from the sky.

A 50-mm focal length lens is adequate, and a telephoto lens is useful for close-up detail at greater distances. Use a tripod for telephotos and in low light conditions or strong winds, to reduce shake. A wide angle lens is useful for single frame views. Polarizing, skylight, or ultra-violet filters help to cut down interference from atmospheric haze.

Colour print film should be used if a suitable photocopier is available to reproduce enlargements. Film speeds of 100 or 200 ASA normally permit sufficient detail to be recorded; carry a 400 ASA film in case daylight deteriorates, though the coarser grain may lose some detail.

Technique

All relevant viewpoints should be recorded and the most important selected for design. If the landscape composition is too wide to be encompassed in a single frame, a number are taken and subsequently joined. Short focal length lenses cause distortion between adjoining frames which makes this difficult. Distortion is minimized with focal lengths of 135 mm and over. Successive exposures are overlapped by about 25 per cent when taking composite panoramas. A tripod ensures accurate horizontal alignment and avoids camera shake.

Exposure is set on that part of the landscape away from the sky and of comparabe tone to the subject of the view. This avoids under-exposure of the fore- and middle-ground. Weather, quality of light, and direction of sun affect results; clear air and bright light from an overcast sky are ideal. Bright sunshine can give too much contrast and loss of detail in shadows, although this is less serious when the landscape is top or front lit. Sunny conditions are acceptable as long as too much detail is not lost. Patchy cloud affects exposure, particularly of panoramic composites, hiding detail as it passes across the view. Avoid hazy or misty conditions, especially for long views.

Each viewpoint should be recorded on a map and the location noted on the negative and 6 × 4 inch prints. Enlargements up to 8 × 6 or 10 × 8 inches are easier to use, with 7 × 5 inches for subsidiary views.

Photographs should be taken from main public view points related to roads, habitation, etc. It is often useful to have an overall view, taken at a high level, which can be used in conjunction with others to co-ordinate the broad design.

(a)

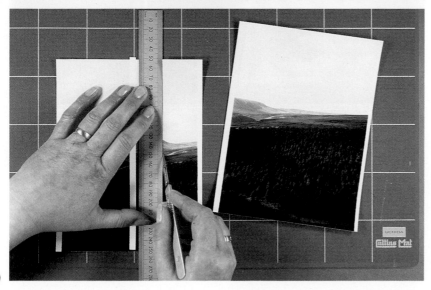

(b)

Fig. 16.1. (a) Identify area of overlap and the portion which gives the best coincidence of features of the landscape. (b) Cut the first print with sharp knife or scalpel within the limits of overlap where there is a good match of line and colour. (c) (see p. 338) Overlap that print accurately on its neighbour, and tape temporarily and securely in position. (d) (see p. 338) Cut through both prints close to the first cut. Press hard on the straight edge to ensure that the prints stay in position once the tape is cut. (e) (see p. 339) Remove surplus, butt accurately in position, and tape temporarily on the front of the prints. Then tape the join permanently on the back, rubbing down for good adhesion. Remove tape from the front of the prints. (f) (see p. 339) Repeat for subsequent frames and trim edges of completed panorama top and bottom.

16.1 (c)

16.1(d)

16.1 (e)

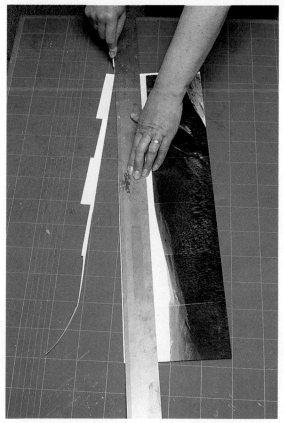

16.1 (f)

Aerial photographs

Aerial photographs or perspectives with acetate overlays are a most accurate aid to setting out on the ground, especially with afforestation. They also assist interpretation of ground level photographs during design work. They are seldom sufficiently accurate for measurement of areas; this is best done from maps.

Aerial photographs of woodland should be as up-to-date as possible. This is less important on open ground which usually changes little over long periods of time. If possible, the scale of aerial photographs should be that of the main maps used for design, usually 1:10 000. Stereo pairs of photographs are useful to give an impression of the three dimensions of landform, and can be used for drawing visual forces which can then be transferred easily to photosketches. Note that the exaggeration of vertical scale as seen through a stereoscope can be misleading. With experience, stereo pairs of photographs can even be used for design related to landform if ground level photographs are not available, but only as a last resort.

Maps

A 1:50 000 map should be used for a broad assessment of the landscape and for planning the landscape survey. More detailed survey and the landscape design require a 1:10 000 map with contours. This is the base map for area measurement and accurate layout of forest shapes.

The landscape survey map is a working document on which necessary information is recorded clearly. It should, therefore, be neat, with different subjects such as elements of diversity, characteristic features, aesthetic problems, and so on, recorded in different colours. Important information can be given greater weight of line or stronger colour. Record visual forces in landform on a separate contour map.

Where the variety and density of information becomes confusing, 'sieve maps' can be used, in the form of clear acetate overlays on a contour map. These enable different classes of information to be displayed in a variety of combinations, and allow potential conflicts and opportunities for co-ordinating different interests to be identified.

Maps of designs should be accurate and neat, with forest shapes outlined in fine black lines and colour coded on the same basis as the relevant photosketches.

Topographic sections

Topographic sections are useful for checking the visibility of trees in relation to screening landform and scale of open space close to the skyline. Contours on many maps are imprecise and a margin of safety is necessary to ensure that unwanted small areas of open ground or narrow tree belts are not visible. The vertical scale of the section can be exaggerated for more accurate measurement, but where the section is to be used for illustrative purposes the exaggeration should not exceed 30 per cent.

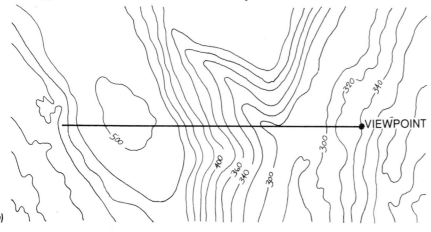

(a)

Fig. 16.2. Making topographic sections from a contour map. (a) Draw line of section on map from viewpoint across the relevant part of the terrain. Project lines at right angles from where the line of section cuts the contour to the corresponding heights on the sectional drawing above. (b) Connect the points on the section and draw line(s) of sight from the viewpoint. Illustrate tree heights at appropriate age intervals as required.

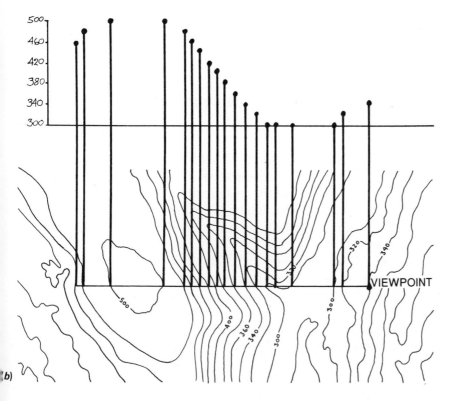

(b)

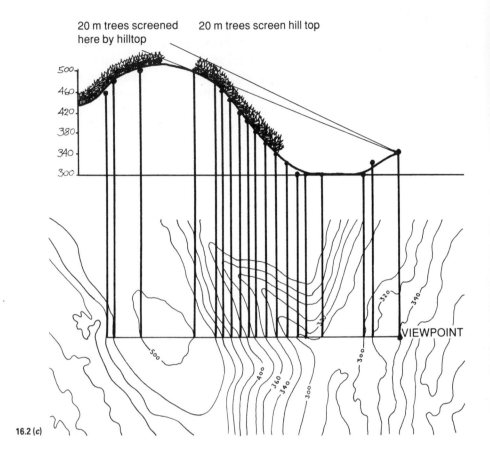

16.2 (c)

Drawing and illustrative techniques using photosketches

Illustrations of landscape design should be as realistic as possible in the time available. The contrast between different elements of the forest should be slightly exaggerated to encourage careful shaping; vague illustrations which play down the visual impact of the forest encourage bad design and reduce the credibility of the design process.

Conifers should be shown by hatching of steeply angled texture and that representing broadleaves given a more rounded shape. The colours in Table 16.2 have been found to give best results.

When showing new planting on the photosketch, represent forest shapes as if at ground level; it is then easier to relate them to maps and aerial photographs. Where 15–20-year-old trees are

TABLE 16.2

	Summer views	Winter colours
Conifers	dark green (all species)	
larch		orange
Scots pine, Sitka spruce		blue green
Norway spruce, Douglas fir, Lodgepole pine		dark green
Broadleaves	lime green	mid-brown
Felled areas	white	white
Young trees, 5–10 yrs	light olive	usually shown in summer colours
Young trees, 10–15 yrs	spring green	

(a)

(b)

Fig. 16.3. Progress of illustration. (*a*) Draw forest shape in pencil. (*b*) Allowing for tree height, add an irregular canopy outline for conifers, and more rounded outline for broadleaves. (*c*) (see p. 344) Fill in conifer areas with zig-zag strokes, working horizontally across the drawing and overlapping successive layers slightly. (*d*) (see p. 344) Completed area showing conifers and broadleaves.

to be shown, sufficient height is then added **upwards** from this line, and an appropriately irregular canopy outline drawn in coloured pencil. The texture of the forest area can then be filled in.

When illustrating felling, forest shapes should be drawn at canopy height, which is easier to relate to the map or aerial photograph. Tree height is then projected **downwards** and the revealed edge has to be illustrated. The canopy can then be shown as described above.

Take time to apply forest texture neatly, with sharp pencils, so that a good design is not devalued because of poor graphics. When applying two colours, as when illustrating mixtures, apply the lighter colour first as this allows the darker to be applied more freely without smudging.

(c)

(d)

(a)

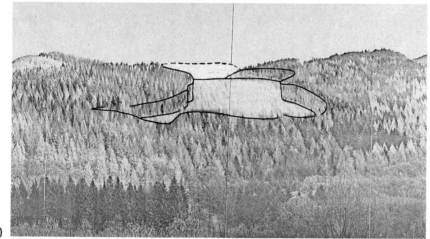

(b)

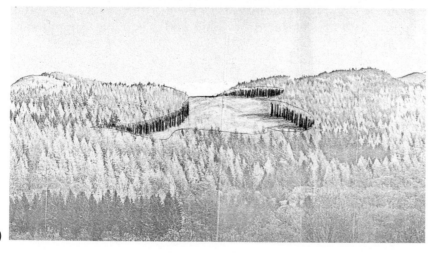

(c)

Fig. 16.4. Illustrating felling coupe. (a) Outline of felling shape is drawn in pencil at canopy level. (b) The extent of tree height related to stand ages and ground level is outlined, and any revealed skylines. The tree texture is erased from the felling area and coloured white. (c) The revealed trunks are shown as black vertical lines and shadows added at ground level in grey. Tree crowns seen in elevation can then be coloured densely in green and the canopy filled in as described for planting.

Forest shapes should first be drawn in pencil so that they can be easily altered. Alternatively, tracing paper can be overlaid to allow revisions without losing the original design. If the revision is an improvement it can be traced from the tracing paper to the photo-sketch using a light box or taping the tracing to a window. More substantial amendments can be made by cutting out and replacing sections of photosketch.

Transfer of design shapes

The process of transferring design shapes from one view to another, and to aerial photographs and maps is essentially the recording of known reference points and identifiable features between two views and interpolating shapes in between. This involves some trial and error, and is made easier if acetate overlays are used on photographs rather than working directly from one photosketch to another. Aerial photographs or a view from a higher level often help this process. First identify as many points as possible along the edge of the shape. These may be natural features, crags, broad-leaved trees, or else points at which the shape crosses forest roads or rides. Then sketch in as much of the shape as possible between these points.

A similar procedure is used to transfer a shape from an aerial photograph to a map.

The work of assessing a design from a variety of viewpoints is greatly speeded up by the use of computer landscape design pro-grammes, where the transfer of shapes between views and between perspective sketches and maps is possible. The development of such programmes continues to be rapid, and it is well worth examining the possibilities of using them if there is a substantial amount of design work to be done and a suitable computer is available.

Methods of laying out landscape designs for new planting

Forest shapes have such an effect on the landscape that accurate mapping and layout on the ground is essential. Unless this is done properly, all previous design work can be wasted. Layout by all methods should be pegged out on the ground before cultivation (except D) and certainly before planting.

Method A: layout of planting direct from sketches from a distant viewpoint with the aid of radio

A member of the project team, familiar with both principles of design and the particular site, stations himself as controller at successive positions where sketches were made. He directs a 'flagman' on the site by radio along the boundaries of the shapes indicated on the sketches, using field glasses. The flagman wears bright coloured clothing or carries a large white or yellow flag so as to be easily seen and directs his assistant(s) who mark the boundary with markers visible to the controller. The latter must be able to see the developing shapes of the boundaries.

This method overcomes the difficulties of reconciling shapes on a map with reality on the ground, particularly in broken hilly country. If the map is of doubtful accuracy, this method may ultimately be less time consuming than other methods.

Method B: layout of planting direct from sketches without radio

Where radio is unavailable or ineffective, techniques similar to Method A can be used. The flagman is briefed initially from the main viewpoints, using an annotated sketch or map showing visible features. Once marking is completed, adjustments are identified from the viewpoints and implemented on site by means of additional notes on the sketch.

Method C: layout of planting direct from landscape design map or aerial photograph

Method C is similar to Method A; the technique is used to adjust forest shapes previously marked on the ground from a map or aerial photograph with overlay. If the design has been plotted accurately on the landscape map, the amount of adjustment should be small.

Method D: layout of planting directly by ploughing

This method can be used where slopes can be ploughed or otherwise cultivated in any direction, with an experienced operator and where the map is accurate. The layout is first pegged minimally from a 1:10 000 landscape design map, then outlined using a plough or similar cultivator to indicate boundary shapes. Important adjustments can be made later from sketches with the aid of radio as in Method A. This method is the least accurate, but provided there are few adjustments there are significant savings in time.

Method of layout of felling coupes

The layout of felling coupes is more difficult because there are often fewer features which are visible in the distant view and on the site. Time can be saved by preparation of information on the map. As many known points as possible, along the boundary of the coupe or nearby, should be identified on the map, such as forest road and ride junctions, or other visible features. The remainder of the coupe boundary can be located by compass bearing from the known points, and curved shapes interpolated. The means of recognizing known points on the edge of the coupe should be noted on the map in advance. Fluorescent tape placed around trees marks the boundary effectively.

If there is doubt about the location of the coupe boundary at any point, play safe by marking the limit of felling about 30 m inside that designed on the landscape plan. Adjustments can be made while felling is in progress using Methods A or B above.

The methods described above may seem laborious, but it is essential to get shapes right. As a sound pattern of coupe shapes develops, there are fewer boundaries to set out because adjoining shapes are defined by earlier coupes.

Fig. 16.5. Progressive setting out of coupes. (*a*) Map of the felling coupe to be set out. (*b*) Known points (KP) and offsets (OS) are identified along the edge of the coupe. A line 30 m inside the coupe edge is located along stretches of the boundary where there are few known points or offsets. (*c*) The boundary is set out and marked on site, taking account of any 'safety margin' by means of compass bearings and measurement on the ground. Curved shapes are then superimposed on straight sections of boundary.

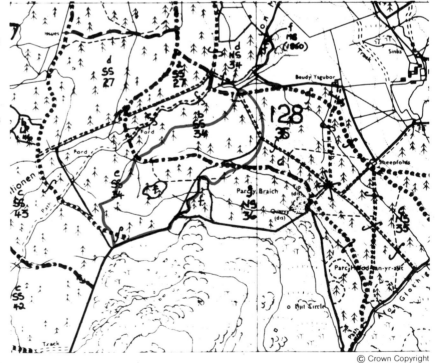

(*a*)

© Crown Copyright

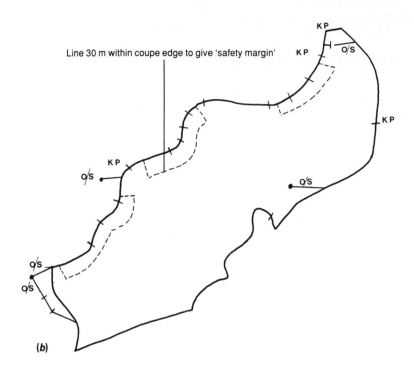

Line 30 m within coupe edge to give 'safety margin'

(b)

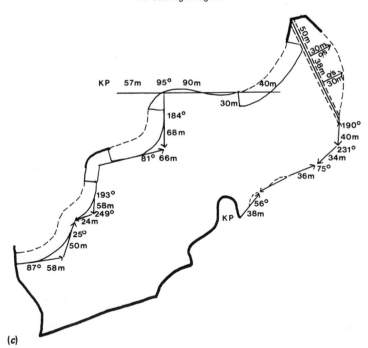

All bearings magnetic

(c)

17 Management of landscape design projects

- Landscape design should be carried out well in advance of any operation which has visible effects.

- The opportunity to improve the landscape should be considered in the planning of all such operations.

- Landscape plans should be prepared by teams led by the local line manager and including one person solely responsible for design.

- The project programme, procedure, and the individual responsibilities within it should be clearly understood by all team members.

The success of a forest landscape plan depends as much on its management as on the quality of design. In addition to the design objective of achieving a satisfactory balance between forest production and landscape is the need to demonstrate that the forest industry genuinely cares for, and can create, attractive landscapes. The publicity given to a project can be as important as the timing, smooth-running implementation, and monitoring.

The forest manager must play a leading role because designs imposed from outside rarely succeed. He should guide the project and must establish a constructive relationship with the designer, who is responsible for much of the detailed work. The manager must be sufficiently aware of the effects of forest operations in the landscape to initiate major projects, take opportunities to make landscape improvements and ensure that plans are implemented during operations, and to recognize unforeseen difficulties as they arise later.

Programming landscape design with forest operations

Landscape design is most effective if it is considered in the earliest stages of planning. It should be borne in mind when the purchase of land for afforestation is contemplated because of the broader issues outlined in Chapter 4.

Design work for new planting must be completed before fence lines are set out or ground preparation begins. Species layout should obviously be agreed before the supply of plants is arranged.

The design of woodland along public roadsides and footpaths should ideally be part of any planting or felling design, but it can be carried out at any stage. In wind-susceptible stands it should take place as early a possible up to the time of first thinning; thereafter, it is best postponed until nearer final felling.

Felling design in thinned crops should if possible be started about 10 years before felling starts so that limited areas can be felled prematurely to develop greater age diversity. A complete felling and restocking plan should be agreed before operations begin.

The close integration of landscape improvements with forest operations ensures that no opportunities are inadvertently missed. The Table 17.1 shows the design work required and opportunities for improvement associated with successive forest operations.

Project management

The project team

The demands placed upon forests are now so diverse that no individual can expect to be expert in all the disciplines required. Every manager must, however, understand the main issues concerning forest production, landscape, wildlife, recreation, archaeology, and hydrology. It requires a team headed by the local manager and advised by a landscape designer to produce an attractive and functional landscape plan.

Everyone in the team must be committed to producing a plan which balances all the issues. It is the responsibility of the project leader to ensure that this is understood, and to create a spirit of integrated teamwork. Any member of the team who is subsequently dissatisfied with the results should only be able to say 'We were wrong' and not 'They were wrong'.

The manager is responsible for ensuring that the design can be efficiently managed, while the designer must draw up a plan to take account of all the issues and ensure satisfactory appearance. The best results are obtained when each suggests solutions to the other's problems. It is often easier for the project leader to make a balanced judgement of the design if the operational and economic aspects are assessed by an assistant.

TABLE 17.1 Forest operations, programming of design, and opportunities for landscape improvement.

Operation	Chapter	Programming of design for operations	Landscape improvement opportunities	Comments
Cultivation	11,	Planting design should be completed, at least for external margins and open spaces, before this operation begins		
	11,	Ploughing design on gentle slopes	Varying spacing and direction of furrows in edge, especially beside roads, paths and recreation areas	
Scarification	11,	"	As above on planting sites	
Preplanting herbicide	12	"		Bands may appear intrusive so complete or spot application preferred in sensitive landscapes
Drainage	11,	"	"	
Planting	6, 7, 8, 9, 10,	Complete design should be agreed far enough in advance to be accurately pegged. Ideally, it should be initiated before land purchase		
Beating up	8		Varying crop edge by not beating up within 5 m	Possible habitat benefits
Fertilizer application	8		Creating varied growth by leaving edges untreated	"
Deer control	9	Management areas should be included in the planting plan and always as part of an overall pattern of open space	Additional control areas can provide more diversity of open space within the forest Limited browsing may be accepted to vary size in visible edges	"

Operation				
Weeding	8, 9, 13	Plans for roads, paths and recreation areas needed in advance (+ conservation rides) preferably at time of planting	Beneficial groups of broadleaved species can be left untreated along visible edges e.g. beside roads, paths and recreation areas	This should be limited to groups which provide identifiable visual or ecological benefits and not used to justify neglect
Cleaning		"	Consider removing conifer around desirable groups of broadleaved along visible edges	"
				Could be extended further for wildlife conservation
Respacing	8, 13	"	Consider reshaping all edges against open ground or routes and additional increases in spacing at the edge	Where original planting design is now considered inadequate
		Improvements to margins against open space need to have been designed. Coupe shapes in non-thin areas should have been identified in advance	Make severance cuts along improved lines of external margins and future coupe edges in non-thin areas so as to establish stable, green edges ready for subsequent felling	Tops should be removed from areas which are to be left open; consider method and cost
				Where original planting design is now considered inadequate
Road construction	11		Improve the space around roads used for walking by clearing and retaining vegetation selectively at roadside	
Roads maintenance			"	
Fence maintenance and replacement	11, 13	Refer to existing roadside plans	Remove unnecessary fencing from recreation sites and public roadsides; replace behind non-vulnerable trees if required	

Operation	Chapter	Programming of design for operations	Landscape improvement opportunities	Comments
Brashing	8, 13	Overall plans for roadsides, path-sides and recreation landscapes should preferably be prepared in advance but are not essential	Treat external edges selectivity and those beside roads, paths, and recreation areas	High pruning may also be necessary to achieve sufficient variety
Firebreaks	10, 11	Incorporate alignment into planting plans if possible	Improve appearance and alignment of breaks in sensitive views. Substitute other methods in places e.g. controlled burning or belts of 'fireproof' species	
Protective burning			See fire breaks	Consider shapes
Early thinning	11	Attention to design of rack alignment may be needed in sensitive landscapes	Opening up features and improvement of forest shapes by clearing and not replanting	Provides an opportunity to meet additional demand for small roundwood; vulnerability to windthrow (see respacing, above) may influence timing; risk is probably lower if this operation is carried out before first thinning, but so is size of produce; worker safety is a prime consideration in rack layout
		Areas of thinning for landscape reasons identified in non-thin areas	Improvement of forest edges beside roads, paths, and recreation areas	Consider brashing and high pruning as well
		Coupe boundaries in non-thin areas should be designed in advance	Last likely opportunity to make severance cuts in non-thin areas	
			Removal of conifer from key broadleaved areas	Broadleaved areas must provide specific environmental benefit

Late thinnings	11, 12, 13	Initiate felling and replanting design so that areas of premature felling can be identified ready to exploit marketing opportunities	Identify limited areas of premature felling close to roads, paths and recreation areas to diversify landscape, screen subsequent felling and maintain internal scale; thin edges of future retentions in advance	To allow greater stability to develop before the edge is exposed
Scrub clearance	12	Following design of felling coupe	Identify coherent groups of well-formed broadleaves which might be retained to diversify the space of a felling coupe	See section on 'Overhead cover'
Felling	12	Felling and replanting design should be complete before felling starts	Improvement of all design; thinning of coupe edges	
Replanting	10, 12	Check species prescribed in landscape plan and revised layout of open spaces	Delayed replanting on some areas can be used to recover felling design lost as a result of windthrow; plant up all geometric rides	These are essential measures to reduce highly intrusive effects

Roles and programme in the design project

The person who has the final say in acceptance or otherwise of a design may, for convenience, be referred to as the client. He provides the will to achieve a balanced design, and the power to approve it and ensure its implementation. He should be well informed on environmental issues and have a clear idea of landscape standards.

The role of the project leader is to motivate and co-ordinate the project team. He is responsible for the successful completion of the project and may be required to adjudicate on unresolvable conflicts between the requirements of forest operations, landscape, recreation, and wildlife. He ensures the progress and co-ordination of the project, obtaining necessary advice from other specialists. He needs an awareness of landscape design and other environmental matters, preferably as a result of training.

The designer provides a service by carrying out the bulk of the design work and must be prepared to advocate good standards of design. He should warn the project leader if he thinks a particular decision will result in an unacceptable appearance, and demonstrate it by illustration if necessary.

The following diagram illustrates the sequence of stages of the project and the responsibilities of those involved.

Briefing

Briefing is an exchange of views and information between members of the project team, not an imposition of one person's views upon the others. It is important that the project leader should not constrain the designer unnecessarily at this stage.

A productive briefing meeting needs careful preparation. The project leader should consider which other specialist(s), e.g. in ecology or recreation, might usefully contribute to the project; it is better to have extra people initially than to give separate briefings. If the staff who will be responsible for implementing the design are present at the briefing meeting, their commitment is likely to be greater.

The briefing meeting should be used to inform the team of the landscape and management issues, and to discuss how the project will be conducted. Each team member should be made aware of his individual and collective responsibilities within the team. If any members of the team are unfamiliar with landscape design it is often useful if the designer explains the principles and practice, using examples of previous work.

The designer should be advised of the needs and objectives of other disciplines so that they can be incorporated into the design. This may happen after the briefing meeting, but preferably before sketch designs are produced, and certainly before a design is agreed. Publicity and outside consultation should also be discussed where appropriate and its co-ordination with the project agreed.

The designer must be free to question managers in an uninhibited way so that all possibilities of design can be explored. This requires tact to ensure that questions are not interpreted as criticism. The design should try to bring out the less tangible expectations of the rest of the team, in terms of quality, attention to detail, timescale of the project, etc. These can significantly affect the co-operation of team members.

The designer should ensure that he has obtained or arranged for all the information required from managers. Following the briefing meeting and landscape survey, the designer should prepare a programme with target dates for each stage of the project, for approval by the project leader.

Changing designs

A design may have to be amended following discussion in the project team or to gain approval of the client. It is vital that the designer should be willing to modify his work and that other team members are constructive in suggesting changes. The design must not be treated as something fixed, for the manager to simply accept or reject, but as a path to a solution for implementation on site. The same applies when windthrow or other events impose changes on a design during implementation, and the design of certain areas has to be re-done. This does not mean that the original approach was wrong. There is no single correct design for a landscape, just a range of good and bad solutions. Monitoring and review are, therefore, important aspects of landscape plans and there is no harm in making sensible modifications within the spirit of the original concept.

HOUGHALL
COLLEGE
LIBRARY

Economic appraisal of landscape plans

Accepting the difficulties of placing a cash value on landscape design, it is worthwhile assessing the costs of treatment above a basic level of design, and comparing them with the aesthetic improvement and the average figure for the whole forest enterprise.

The basic design level should not be mere avoidance of intru-

PROJECT TEAM

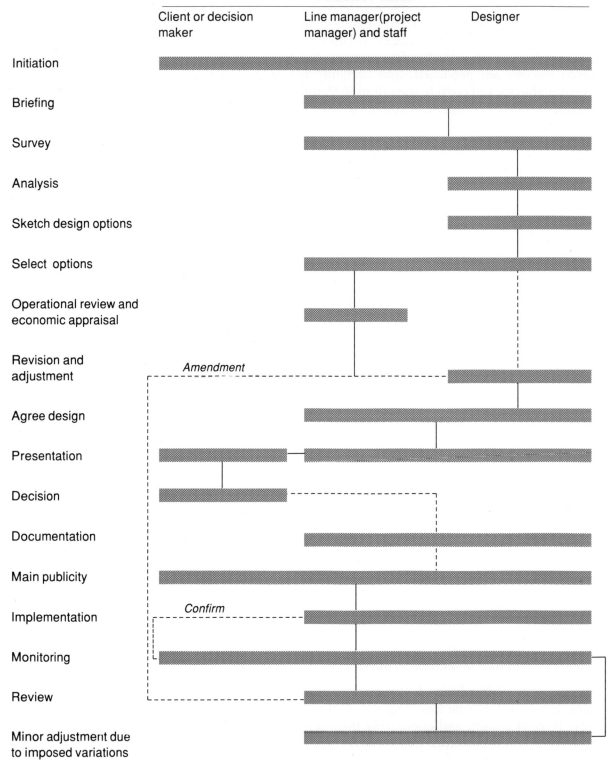

Decision maker must agree initiation of project and also, in the project team the sensitivity of the landscape and the level of timing and consultation he requires.

Briefing is a bilateral discussion between project manager and designer which develop the detailed basis and objectives for design and sets deadlines. Unrealistic and unnecessary constraints should not be set. Consider publicity and consultation.

It is easier for the line manager (+ staff) to gather forestry information for the designer who carries out landscape survey and records other aspect of the design project.

Unsatisfactory options may be needed to demonstrate problems or assess costs.

If a major revision is needed project team members may need to help the designer generate new options.

The designer should be present, to explain the aesthetic issues and answer questions.

At least one map for each office and spares for field use, etc.

Designer to provide necessary graphic expertise and explain the design if required.

The designer should monitor accuracy of implementation and aesthetic quality.

The designer should be consulted on the aesthetic issues whenever any adjustments are necessary.

Project team

sion; best value for money rarely occurs at this point. Additional measures related to landscape sensitivity are invariably required to co-ordinate the design to the point where the quality of the total effect is maximized. Different parts of the design interact to such an extent that a comprehensive design is always necessary; otherwise, successive measures, designed and costed separately, may allow excessive expenditure to accumulate over a period of years in an uncontrolled way. Any economic appraisal should, therefore, be made for a complete set of measures before implementation begins on a particular area.

The simplest area to use for this type of appraisal is the discrete woodland area, to the extent of all its boundaries with open land. Where there are a number of woods visible in the same key views, it is better to group them together and cost the whole landscape composition. Larger forests may extend beyond a single composition, e.g. over two sides of a hill. It is then advisable to divide the area according to the most dominant visual features (skylines, ridgelines, separate landforms) to make both design and appraisal more manageable, bearing in mind that these are artificial divisions across which design and consequential costs interact.

The costs to be assessed are those in excess of the basic standard of design required to allow forestry to proceed in sympathy with the landscape. These costs are attributable mainly to reductions in potential timber revenue because certain land is left unplanted, less productive species used or felling at ages other than the economic optimum. There may also be small increases in operating costs where work is done purely for landscape reasons. The method of costing will depend on the financial objectives and accounting conventions of individual forest enterprises, but the following table gives the main elements to be assessed.

In judging the costs and benefits of a landscape design it is essential to balance costs against the aesthetic benefits as illustrated in perspective drawings and in the light of the sensitivity of the landscape. Costs expressed as a percentage of discounted revenue avoid confusion due to differences in productivity on different sites and make comparisons easier.

There may be small areas of unavoidable high cost which are necessary to eliminate a particular point of visual intrusion or to achieve an overall effect. At first sight they may seem to be expensive details, but their cost has to be weighed against the the visual consequences of not carrying them out.

One may have to choose between several design options. In doing so it is extremely useful to have an indication of the economic consequences of each. There may well be cases where a slight improvement in the design is only attainable at considerable cost; or where a small additional expenditure raises a design from satisfactory to excellent. Either way, it is only fair to the client to spell out the net costs associated with each option. Only with this information is he in a position to take a balanced view of the costs and benefits of the design proposals.

Consultation with statutory authorities

In Great Britain the Forestry Commission regulates tree felling, and provides grant aid to private forestry for planting and re-stocking. When considering applications for felling permission or planting grants, the Commission consults agricultural departments, local planning authorities, and other statutory authorities as appropriate, to ensure that requirements of land use, agriculture, amenity, recreation, nature conservation, and archaeology are taken into account before decisions are reached. The views of the Countryside Commissions and local planning authorities on landscape sensitivity are an important element in this procedure. Information on the local authorities' views is contained in published structure or local plans, and in National Park Plans.

Implementation of designs

Once a balanced design has been agreed it should be rigorously implemented. Shape being so important, the planned layout should be followed as closely as possible. Any expedient adjustments on the ground are likely to shift the balance away from the design standard required. The person setting out the boundaries should, therefore, be committed to the design and should preferably have been a member of the project team. The agreed design should be accurately recorded on a map which is kept in a safe place. The designer should provide enough copies so that tracing by managers, which can introduce errors, is unnecessary.

Any adjustments shown to be necessary in setting out should be critically examined in the longer view, because even slight changes can introduce symmetry. Canopy screening makes the layout of felling coupes particularly difficult, so those setting out coupe edges should have an understanding of the basic principles of forest shaping. The layout of each coupe should be assessed during felling, in case the layout or design needs improving. When final fellings are sold standing it is vital that the requirement to fell to the marked boundaries is written into the sale contract, with penalties for non-compliance.

The project leader should inform local supervisors of measures to be planned and carried out as part of felling and restocking operations, such as thinning of retained edges, retention or felling of remaining broadleaves within coupe boundaries, varied spacing of plants in edges, merged species margins, areas for delayed restocking to improve forest shapes, and any straight rides which are to be planted during restocking or realigned along coupe boundaries.

Publicity for landscape plans

It may be necessary to consider publicity as an integral part of the landscape planning process, initiated at the same time as the project. Landscape changes are sometimes viewed with concern by the public, often because they have no idea what is going on, and it can be well worth while revealing proposals publicly to forestall ill-informed criticism.

The landscape plan may be publicized by radio, television, or the press, in association with other news or feature items related to the forest. These media are useful in awakening interest in a plan, which can be promoted by a public 'launch' and open day, or other events, such as photographic competitions, guided walks, etc.

Many people find it difficult to understand maps and will only read limited amounts of text, so illustrations showing the future appearance of the landscape are essential to demonstrate good practice. For the non-professional it is a well to have a comparison either with the existing landscape (on site) or with an alternative showing bad practice.

As work progresses the results can be compared with illustrative boards on site showing the relevant part of the landscape plan and photographs of the previous appearance. The comparisons have to be simple, clear, and accurate, with text kept to a minimum. Such displays can be located along a trail demonstrating the implementation of the plan from different viewpoints.

The standard of graphics and materials will vary. Those used in magazines and visitor centres should be of comparable quality and design to the other displays. It may be necessary to commission an illustrator.

Working drawings are more acceptable for temporary exhibitions or talks on site. Permanent on-site illustrations should be protected against weather and mechanical damage.

The best medium for explaining the detail of a landscape plan is an indoor talk with illustrations or slide projector. Outdoor talks have the benefit of reality and potentially a more enjoyable environment, but they are vulnerable to weather. The number of images which can be shown is limited, requiring enlarged working drawings mounted on lightweight board with transparent film applied to the face to protect them from rain. The number who can see, hear, and take part in the proceedings is restricted if weather conditions are poor.

TABLE 17.2 Elements of cost of forest landscape design

Element of cost or reduced revenue	Remarks
Land bought, but left unplanted	Establishment costs saved
	Areas which would not be planted for non-aesthetic reasons should not be included, e.g. transmission lines, deer glades, stream sides
	Some land may be put to other uses, e.g. let or sold
Planting less productive species	Use the difference in discounted revenue between species used and the most productive species practicable on that particular site
Premature felling	Reduced timber revenue is partially offset by discounting; compare with the economic optimum
Delayed felling	Possibly increased timber revenue, but may be offset by discounting
Increased thinning and spacing in edges	Reduction in timber quality because of heavier branching; possible loss of volume
Cultivation	Costs may be greater where more frequent changes of direction are required
Additional fencing	Net increases should be assessed
Maintenance of unstocked land	Unnecessary where fenced out
	May be only partially treated, e.g. to disguise differences in vegetation due to differential grazing. Costs may be reduced by grazing these areas.
Clearance of brash	Limited to felled areas close to roads, paths, etc., or which will not be restocked
Reseeding felled areas to be left unstocked.	Only in most sensitive landscapes; may be necessary where felling cannot follow acceptable shapes because of windthrow risk

Additional weeding on delayed restocked areas	
Respacing, brashing or pruning of edges	
Additional protection costs	Due to selection of more vulnerable or slower growing species
Marking for felling	Setting out coupe edges and marking additional thinnings along them
Additional harvesting costs	No real additional cost where produce is extracted through retained stands by well-selected routes; some increase in costs may arise if logging operations are done in two or more stages instead of one
Felling unsightly whips and scrub on felled areas	Usually only felling is required; sell them standing as firewood?

Glossary

The meanings assigned to terms used in this book are given below.

Aesthetic(s)
(a) The science or study of beauty. (b) The theory or understanding of the perception of the environment by all the senses.

Appearance
The perceived nature of the object or landscape as distinct from its known nature (see *Illusion*).

Asymmetry
The visual balance of dissimilar elements in a design or landscape.

Background
(a) The more distant part of the landscape seen in a specific view, usually beyond about 5–8 km distant, and with only landform and the pattern of woodland and open space visible. (b) The area around or behind an element, providing harmony or contrast.

Back-lit
An object or landscape which is seen with the sun opposite the viewer is back-lit. This causes reflection from more horizontal surfaces while vertical elements are seen in silhouette or shadow and tree foliage appears translucent. Details are obscured.

Balance
A state in which no change to a design or landscape, in terms of adding, removing or moving any element appears to be necessary; when the dynamic influences of all the parts of a composition appear to be in equilibrium. Visual balance may be symmetrical or asymmetric.

Basic elements
The landscape can be defined in terms of basic visual elements of point, line, plane, and volume, all of which can interact.

Beauty
The harmonious relationship of seen parts which, brought together in a composition, give great pleasure to the senses.

Canopied
Covered or enclosed overhead by branches or foliage of trees, while substantially open beneath and often bounded at the sides by a space.

Character
The distinguishing aspects of an element or a landscape; no value or quality is implied and character is usually widely distributed in the landscape being described.

Characteristic

A widely repeated or distributed element of a landscape which is distinctive in itself or contributes to the landscape character.

Closure

The partial enclosing of an area or volume of landscape thereby suggesting a larger element.

Colour

The characteristic of a material which affects the wavelength, quality, and intensity of light reflected from it.

Composition

The collection of different visual elements into a satisfying and identifiable whole. Landscape compositions may exist as natural units, e.g. a hill or valley, without any conscious act of design being involved.

Contrast

The visible differences between two parts placed close together. The greater the visual differences and the closer they are placed, the greater is the degree of contrast.

Context

The specific character and qualities of the surroundings to which an element may be compared, or the recent past as the background to change. It may mean more than the immediately visible surroundings and include the region beyond (see *Background*).

Coupe

A discrete and recognizable area on which felling takes place as a single continuous operation (clear felling) or in sequential stages (shelterwood or group felling). A felling coupe would normally, though not invariably, be restocked as a single entity.

Design

A creative project in which the visual and physical parts are assembled in order to achieve a specific end result. A good design is one in which the parts and their relationship are well balanced and unified in order to achieve the objective as attractively and efficiently as possible.

Disruptive

Destructive of visual unity; tending to separate the elements of a design or detracting from the factors that make them seem part of an overall composition.

Diversity

The degree and number of differences in a composition; an important principle of design is to achieve sufficient diversity to create interest without causing loss of unity.

Direction

The quality of an image which draws the attention from one part of a composition to another in a predictable way.

Edge

The boundary between one part of a landscape and another, whose quality is determined by detail of interval, grouping, size, etc.; the detailed elements of a margin.

Emphasis

A forcefulness of expression which draws attention to an otherwise insignificant element, variable, or factor.

Enclosed

Appearing surrounded, usually of space; enclosure, an area, or space which is enclosed.

Element

An identifiable part of a composition, e.g. basic element, point, line plane, volume.

Feature

An apparently small element which catches the eye individually against a more extensive background.

Foreground

The nearest part of a view in which individual plants and species are visible; generally extends up to 0.5–1 km distant.

Foreshortened view

A view seen at an oblique angle to the ground which makes horizontal distances seem shorter and vertical elements such as trees appear closer to each other or to overlap; a characteristic of convex slopes.

Form

The three-dimensional equivalent of shape.

Front lit

The appearance of a landscape seen with light coming from behind the viewer. Forms tend to become lost, but colours and details show up clearly.

Genius loci

See *Spirit of place*.

Geometry, geometric shapes

Simplicity of shape derived from that branch of mathematics. Characterized by straight lines, right angles, arcs of circles, etc.

Illusion

A visual impression which belies the physical reality; a misleading appearance. It is most common in landscape and design as an illusion of movement or force.

Interlock

A relationship between two elements where one extends into the other.

Interval

The space between repeated elements. It often interacts with size (see below) and may be regular or irregular.

Intrusion

The quality of an element or factor which appears to stand out to the detriment of a design; a serious visual problem or conflict.

Landform

The three-dimensional shape of the ground.

Line

A basic visual element which is strongly dominated by one dimension. It may be a boundary between two planes, or an extended point, or a series of elements. This is a powerful element in forest design.

Margins

The visual boundaries of the forest which are perceived as shapes; species margin, the boundary between areas of different species; external margin, the boundary between the forest and open ground; coupe margin, the boundary between the standing trees and a felled area.

Mass

Solid volume, usually in contrast to space. The forest is perceived from outside as mass.

Middle ground

The space between the background and the foreground in a landscape or picture, usually from about 1–6 km distant.

Naturalistic

Having the appearance of being natural. In strict terms few landscapes in Britain are natural although many appear to be so. A landscape designed to appear natural is described as naturalistic.

Nearness

The proximity of visual elements such that they appear to be part of a group or composition.

Number

When the number of elements in a design increases, the design becomes more complicated and can become increasingly confused.

Perception

The activity by which the mind interprets what the senses (mainly sight) receive.

Perspective

An illustration or drawing as seen by the human eye with the effect of distance.

Photomontage

A technique of illustration whereby parts of different photographs are joined to create a new image.

Photosketch

A perspective produced by photocopying a photograph and usually applying colour to illustrate particular or new elements in the landscape, e.g. new planting.

Plane

A basic element in which two dimensions dominate a surface. It may be flat, shaped, real, or implied: most commonly, the ground plane in landscape.

Point

A basic element in which all dimensions appear small; position and number are usually important.

Position

The location of an object in relation to its surroundings. Vertical, diagonal, and horizontal are three basic positions. It is a very important variable when dealing with smaller, contrasting elements.

Proportion

The size, quality, or degree of a part in relation to the whole.

Quality of landscape

The degree of excellence of a landscape; high quality is excellent, low quality is poor.

Respacing

The removal of young trees, whether planted or naturally regenerated, before they have reached timber size. The operation is usually carried out before the trees are 2 m tall. It is done to provide the remaining trees with more growing space, either as an alternative to normal thinning on sites with high risk of windthrow or where natural regeneration is over-abundant.

Restocking

The replacement of felled trees by desired species, either by planting, by natural regeneration from mother trees on the site or nearby, or by coppice regrowth from the stools of the previous crop.

Retention

Trees deliberately retained on or adjacent to a felling coupe when similar trees are felled. Areas of retention may vary from substantial stands to groups or individual trees. Retentions may be for landscape reasons, to benefit wildlife, or for the purposes of timber production.

Rhythm

The repetition of similar elements so that they appear part of a whole composition and create an illusion of movement. It provides a means of integrating numerous elements in a design and increasing interest.

Scale

Size in comparison with: (a) the human figure; (b) the landscape; (c) the whole composition, i.e. proportion.

Seasonal changes

Changes due to leaf colour or precipitation throughout the year. What appears to be an acceptable design in summer may look intrusive in autumn.

Sensitivity, landscape

A combined assessment of landscape quality, visibility and recreational use which indicates the standard of design which should be achieved in a particular place.

Shape

The variation in the edge of a plane. It is the most important and evocative variable (see *Geometry*).

Side lit

The appearance of a landscape with light coming from the side. This emphasizes three-dimensional quality and texture.

Similarity

The degree to which separate elements have visual characteristics in common. The more similar they are, the more they are perceived as a unified composition, and they do not have to be identical.

Size

The variation in dimension of various elements. It can be used to introduce diversity. Large size is an important characteristic of mature conifers.

Sketch

An illustration of an object or landscape as it is actually seen.

Sketch design

An outline of approximate proposal for a design.

Skyline

The line where the land and sky appear to meet: it is usually dominated by landform and appears large in scale.

Space

Open volume defined by surrounding elements, e.g. by the ground plane or by the canopy overhead.

Spirit of place

The combination of character, features, quality, space, and associations which creates a unique sense of identity in a location (see *Genius loci*).

Symmetry

The balancing of equal numbers of elements or shapes as mirror images on either side of a median line. It creates designs which are artificial, stable, secure, and static.

Tension

The interaction of visual forces, often causing conflict, but can be used to create greater interest in a design.

Texture

The visual or tactile quality of a surface created by numerous repeated elements. A wide interval between elements creates coarse texture, while a narrow interval creates a finer texture.

Unity

The appearance of wholeness and continuity of a design or landscape, or the organization of collected elements into a clearly identifiable composition: the sense of continuity between an element in the landscape and its background.

Value of landscape

The total worth placed by the public as a whole on a specific landscape; impossible to calculate, but it can be assessed in qualitative emotional and comparative terms.

View

An area of landscape seen from a specific point; viewpoint. The precise place from which a specific view is seen.

Vista

A confined view, often artificially contrived, and usually with a frame of trees.

Visual

Describing an image perceived by the sense of sight.

Visual force or energy

The illusion of movement or potential movement created by a static image.

Volume

A basic element in which all three dimensions are easily seen (see also *Mass, Space*).

References and further reading

The following publications are referred to in this book, or are recommended for further study of particular aspects of landscape appraisal and design.

Anderson, M. A. and Carter, C. I. (1987). Shaping ride sides to benefit wild plants and butterflies. In *in Wildlife management in forests*. ICF.

Appleyard, D., Lynch, K., and Myer, J. R. (1984). *The view from the road*. Massachusetts Institute of Technology, Massachusetts.

Boyd, J. M. (1987). Commercial forests and woods; the nature conservation baseline. *Forestry*, 60, 113–34.

Countryside Commission for Scotland (1978). *Scotland's scenic heritage*. Countryside Commission for Scotland, Battleby, Perth.

Crowe, Dame Sylvia (1978). *The landscape of forests and woods*, Forestry Commission Booklet 44. HMSO, London.

Crowe, Dame Sylvia (1981). *Garden design* (2nd edn). Packard Publishing in association with Thomas Gibson Publishing, London.

Crowe, Dame Sylvia and Mitchell, M. (1988). *The Pattern of Landscape*. Packard Publishing, Chichester.

Davies, P. and Knipe, A. (1984). *A sense of place: sculpture in the landscape*. Ceolfrith Press, Sunderland.

Drabble, M. (1979) *A writer's Britain*. Thames and Hudson, London.

Forestry Commission (1986). *Guidelines for management of broadleaved woodlands*. HMSO, London.

Forestry Commission (1988). *Forests and water: guidelines*. Forestry Commission, Edinburgh.

Garret, L. (1967). *Visual design — a problem solving approach*. Van Nostrand Reinhold Co., New York.

Hardy, T. (1878). *The return of the native*, Macmillan, London.

Hibberd, B. G. (ed.) (1986). *Forestry practice*, Forestry Commission Bulletin 14. HMSO, London.

Kaplan, S. (1973). Cognitive maps; human needs and the designed environment. In: *Environmental design research* (ed. W. F. E. Preiser). Dowden, Hutchinson & Ross, Stroudsberg, Pa.

Litton, R. B., Jr (1968) *Forest landscape description and inventories*, USDA Forest Service Research Paper PSW-49. USDA, Washington DC.

Litton, R. B., Jr (1972). Aesthetic dimensions of the landscape. In *Natural environments; studies in theoretical and applied analysis* (ed. J. V. Krutilla), pp. 262–91. Johns Hopkins, Baltimore.

Loudon, J. C. (1833). *The landscape gardening and landscape architecture of the late Humphry Repton Esq*. London.

Norberg-Schulz, C. (1980). *Genius loci*. Academy Editions.

Peterken, G. F. (1981). *Woodland conservation and management*. Chapman & Hall, London.

Rackham, O. (1976). *Trees and woodlands in the British landscape*. Dent, London.

Rackham, O. (1980). *Ancient woodland*. Arnold, London.

Steele, R. C. (1972). *Wildlife conservation in woodlands*, Forestry Commission Booklet 29. HMSO, London.

Smart, N. and Andrews, J. (1985). *Birds and broadleaves handbook*. Royal Society for the Protection of Birds, London.

Springthorpe, G. D. and Myhill, N. G. (1985). *Wildlife rangers handbook*. Forestry Commission, London.

United States Department of Agriculture, Forest Service (1973) *National forest landscape management* Vol. 1. USDA Forest Service, Washington, DC.

Index